Lecture Notes in Control and Information Sciences

Edited by M. Thoma and A. Wyner

Vol. 81: Stochastic Optimization
Proceedings of the International Conference,
Kiew, 1984
Edited by I. Arkin, A. Shiraev, R. Wets
X, 754 pages, 1986.

Vol. 82: Analysis and Algorithms
of Optimization Problems
Edited by K. Malanowski, K. Mizukami
VIII, 240 pages, 1986.

Vol. 83: Analysis and Optimization
of Systems
Proceedings of the Seventh International
Conference of Analysis and Optimization
of Systems
Antiba, June 26-27, 1986
Edited by A. Bensoussan, J. L. Lions
XVI, 901 pages, 1986.

Vol. 84: System Modelling
and Optimization
Proceedings of the 12th IFIP Conference
Budapest, Hungary, September 2–6, 1985
Edited by A. Prékopa, J. Szelezsán, B. Strazicky
XII, 1046 pages, 1986.

Vol. 85: Stochastic Processes
in Underwater Acoustics
Edited by Charles R. Baker
V, 205 pages, 1986.

Vol. 86: Time Series and
Linear Systems
Edited by Sergio Bittanti
XVII, 243 pages, 1986.

Vol. 87: Recent Advances in
System Modelling and
Optimization
Proceedings of the IFIP-WG 7/1
Working Conference
Santiago, Chile, August 27-31, 1984
Edited by L. Contesse, R. Correa, A. Weintraub
IV, 199 pages, 1987.

Vol. 88: Bruce A. Francis
A Course in H_∞ Control Theory
XI, 156 pages, 1987.

Vol. 88: Bruce A. Francis
A Course in H_∞ Control Theory
X, 150 pages, 1987.
Corrected - 1st printing 1987

Vol. 89: G. K. H. Pang/A. G. J. McFarlane
An Expert System Approach to
Computer-Aided Design
of Multivariable Systems
XII, 223 pages, 1987.

Vol. 90: Singular Perturbations
and Asymptotic Analysis
in Control Systems
Edited by P. Kokotovic,
A. Bensoussan, G. Blankenship
VI, 419 pages, 1987.

Vol. 91 Stochastic Modelling
and Filtering
Proceedings of the IFIP-WG 7/1
Working Conference
Rome, Italy, Decembre 10-14, 1984
Edited by A. Germani
IV, 209 pages, 1987.

Vol. 92: L. T. Grujić, A. A. Martynyuk,
M. Ribbens-Pavella
Large-Scale Systems Stability Under
Structural and Singular Perturbations
XV, 366 pages, 1987.

Vol. 93: K. Malanowski
Stability of Solutions to Convex
Problems of Optimization
IX, 137 pages, 1987.

Vol. 94: H. Krishna
Computational Complexity
of Bilinear Forms
Algebraic Coding Theory and
Applications to Digital
Communication Systems
XVIII, 166 pages, 1987.

Vol. 95: Optimal Control
Proceedings of the Conference on
Optimal Control and Variational Calculus
Oberwolfach, West-Germany, June 15-21, 1986
Edited by R. Bulirsch, A. Miele, J. Stoer
and K. H. Well
XII, 321 pages, 1987.

Vol. 96: H. J. Engelbert/W. Schmidt
Stochastic Differential Systems
Proceedings of the IFIP-WG 7/1
Working Conference
Eisenach, GDR, April 6-13, 1986
XII, 381 pages, 1987.

Lecture Notes in Control and Information Sciences

Edited by M. Thoma and A. Wyner

141

S. Gutman

Root Clustering in Parameter Space

Springer-Verlag Berlin Heidelberg GmbH

Author
Prof. Shaul Gutman
Dept. of Mechanical Engineering
Technion – Israel Institute of Technology
Haifa, Israel

ISBN 978-3-540-52361-1 ISBN 978-3-540-46970-4 (eBook)
DOI 10.1007/978-3-540-46970-4

Originally published by Springer-Verlag Berlin Heidelberg New York in 1990.

2161/3020-543210 Printed on acid-free paper.

To my mother

אשת חייל מי ימצא

PREFACE

In 1763 Waring discovered that aperiodicity can be tested using the minus square of the root-pair-difference. This discovery marked the beginning of root clustering investigation. The important ideas of stability and later relative stability, introduced in the 19th century, motivated the following general problem: to find a criterion for the inclusion of the eigenvalues of a given matrix in a prescribed algebraic region in the complex plane. The extensive research carried out in recent years, has brought the subject to a reasonable level of maturity; thus I feel it is about time to bridge the 220 years with an appropriate book. This, to the best of my knowledge, is the first book dealing with the general root clustering problem.

The book covers five basic topics: First - a review of classical results; second - root clustering criteria based on one variable transformation; third - criteria based on composite matrices and polynomials; fourth - criteria based on linear matrix equations; fifth - the image of the criteria in the parameter space, including an application to feedback.

This book should prove valuable for systems and control engineers as well as for mathematicians. In other areas such as physics, the results may be of help in the analysis of stability properties.

Although the book reflects my personal view of the subject, I have included, for completeness, other main approaches to root clustering. The theme of the book is general structures of root clustering criteria, and their image in the parameter space. It is not my intention, however, to replace Routh table, for instance, by a more complicated criterion.

My interest in the subject originated in the pleasant atmosphere of Berkeley, more than a decade ago. I soon realized that there is no contradiction in working with Professor George Leitmann on differential games, min-max, etc., while at the same time being involved with Professor Eli Jury in new ideas about stability. George's devotion to mathematical carefulness and Eli's enthusiasm for stability contributed much to my education. However, most of my research on root clustering has developed during my work at the Technion, thanks to the the friendly spirit of my colleagues in the Department of Mechanical Engineering. Among my past students I wish to mention Dr. Fabien Chojnowski and Dr. Hedi Taub who made a significant contribution to the theory presented in the book. Part of the results in Sections 7.5, 7.6 , and the Appendix are due to my Ph.D. student Mani Fischer. I wish to thank Mrs. R. Alon and Mrs. M. Schreier for typing the manuscript and I. Katner for computer drawings. Last but not least, I wish to thank my wife Yaffa and my children, Rakefet, Oren and Michal who inspired my life and work. Without their patience and support, this work would not have been completed.

תם ונשלם שבח לאל בורא עולם

תשרי התש״ן

Shaul Gutman

Haifa, Israel, Oct. 1989

CONTENTS

LIST OF SYMBOLS

ℵ	a region in the complex plane
cl(ℵ)	the closure of ℵ
$\mathbf{R}$, R	the set of real numbers
$\mathbf{C}$, C	the set of complex numbers
$\mathbf{R}^n$, R^n	n-dimensional real space
$\mathbf{C}^n$, R^n	n-dimensional complex space
$\mathbf{R}[z_1, ..., z_n]$	the set of n-variate real polynomials
$\mathbf{C}[z_1, ..., z_n]$	the set of n-variate complex polynomials
$\times$	Cartesian product
$\in$	a member of
$\cup$	union
$\subset$	is a subject of
$\cap$	intersection
$\forall$	for all
$\exists$	there exists
$\emptyset$	the empty set
$\Leftrightarrow$	if and only if
$\mathbf{R}^{n\times m}$, $\mathsf{R}^{n\times m}$	the set of n×m real matrices
$\mathbf{C}^{n\times m}$, $\mathsf{C}^{n\times m}$	the set of n×m complex matrices
$\bar{a}$	complex conjugate of a
A'	matrix transpose
A^*	matrix conjugate transpose
$\otimes$	Kronecker product
$\odot$	bialternate product
$\cdot$	Schur (term by term) product
$\downarrow$	stacking operator
$\uparrow$	inverse stacking operator
tr	trace of a matrix
σ	spectrum, collection of all eigenvalues (roots)
det	determinant of a matrix
$\vert\ \vert$	absolute value, determinant
$\Vert\ \Vert$	norm
p.d.	positive definite
p.s.d.	positive semidefinite
Coef	all coefficients of a polynomial
$\overset{mod}{=}$	modular equality; equals on the spectrum
Min	minimum
s.t.	subject to

Chapter 1 : INTRODUCTION

The stability of linear differential equations has attracted the attention of scientists since the 19th century. Later, the same subject arose with respect to difference equations. In systems design, however, stability is not sufficient for "good" behavior. To measure the distance from the boundary of stability, we define the concept of ***relative stability*** for a given system, and relative stability in ***parameter space*** for a system whose parameters are to be selected by the designer. We start our discussion by presenting a mathematical model.

1.1 MATHEMATICAL MODEL

Consider the dynamic system S given by :

$$S: \quad \begin{aligned} P(d)q &= Q(d)u \\ y &= R(d)q + S(d)u \end{aligned} \tag{1.1}$$

where $P(\cdot)$, $Q(\cdot)$, $R(\cdot)$ and $S(\cdot)$ are polynomial matrices, $u \in R^m$ is the ***input*** vector, $y \in R^r$ is the ***output*** vector, and $q \in R^p$ is the ***generalized coordinate*** , or the ***partial state*** . For ***continuous***-time system, d is the ***derivative*** operator

$$d \overset{\Delta}{=} \frac{d}{dt} ; \quad d^i \overset{\Delta}{=} \frac{d^i}{dt^i} .$$

For ***discrete***-time systems, d is the ***shift*** operator

$$dy(t) \overset{\Delta}{=} y(t+T), \quad d^k y(t) \overset{\Delta}{=} y(t+kT).$$

In the special case where $P(\cdot)$ is of first order, $P(d) = dI-A$, we say that system S is in ***state-space*** representation. Replacing q by x, we then have :

$$S: \quad \begin{aligned} \dot{x} &= Ax + Bu \\ y &= Cx + Du \end{aligned} \tag{1.2}$$

for continuous-time, and

$$S: \quad \begin{aligned} x_{k+1} &= Ax_k + Bu_k \\ y_k &= Cx_k + Du_k \end{aligned} \tag{1.3}$$

for discrete-time. The vector x is the ***state*** vector. In what follows, we adopt the following dimensions : $x \in R^n$, $u \in R^m$, $y \in R^r$. Thus, $A \in R^{n\times n}$, $B \in R^{n\times m}$, $C \in R^{r\times n}$, and $D \in R^{r\times m}$ The connection between (1.1) and (1.2) is not only for the special case. If the pair $\{P(\cdot), Q(\cdot)\}$ is left coprime, that is, if (1.1) does not have hidden modes, then it is known that S given in (1.1) has a state-space representation of the form (1.2) or (1.3) with minimal order equals $\deg|P(d)|$. The problem of transfering (1.1) into state-space form with minimal dimension is known as ***minimal realization*** and will not be discussed here.Using the Laplace Transform with zero initial conditions, the input-output relation has the form :

$$y(s) = G(s)u(s) \qquad (1.4)$$

where for (1.1), and (1.2)-(1.3), respectively :

$$\text{(i)} \quad G(s) = R(s)P^{-1}(s)Q(s) + S(s)$$
$$(1.5)$$
$$\text{(ii)} \quad G(s) = C(sI-A)^{-1}B + D$$

the matrix G(s) is rational and called ***a transfer matrix.***

1.2 ASYMPTOTIC STABILITY

Asymptotic stability is fundamental in linear dynamical systems. Let us start with a definition with respect to the model (1.2).

Definition 1.1 : The free system $x = Ax$ is ***asymptotic stable*** , if $x \to 0$ asymptotically from all initial conditions.

Applying the Laplace Transform to $\dot{x} = Ax$, we find

$$x(s) = (sI - A)^{-1}x_0 = \frac{1}{\Delta(s)}\operatorname{adj}(sI - A)x_0$$

where $\Delta(s) = |sI-A|$.

Thus, each component of x has the form $\alpha_0(s)/\Delta(s)$, where $\alpha_0(s)$ is a polynomial whose coefficients depend on the initial conditions. Applying partial fraction expansion and taking the inverse Laplace Transform, we obtain (after replacing s by the more frequently used symbol λ)

$$x_i(t) = \sum_{i=1}^{n} \gamma_i t^{k_i} e^{\lambda_i t} ,$$

where λ_i, the roots of $\Delta(\lambda) = |\lambda I-A|$ are called the ***eigenvalues*** of A, and k_i is the multiplicity of λ_i.Using the notation $\lambda_i = \sigma_i + jw_i$, we see that $e^{\lambda t} = e^{jwt}e^{\sigma t}$. Thus, $|e^{\lambda t}| = e^{\sigma t}$. Since $e^{\sigma t}$ dominates t^k, we conclude that $x_i(t) \to 0$, if and only if $\sigma_i < 0 \; \forall i$.The reader can verify the same conclusion for (1.1), where now, $\Delta(\lambda) = |P(\lambda)|$.

Concerning the discrete system (1. 3) or (1. 1), $e^{\lambda t}$ is replaced by λ^k. Thus, for $\lim_{n \to \infty} x(n) = 0$ we require $|\lambda_i| < 1$. As a summary, we define the ***characteristic polynomial*** $\Delta(\lambda)$, by :

(i) $\Delta(\lambda) = |P(\lambda)|$ for (1.1)

(1.6)

(ii) $\Delta(\lambda) = |\lambda I-A|$ for (1.2), (1.3) .

The solutions of $\Delta(\lambda) = 0$ are called the ***roots*** of $\Delta(\lambda) = 0$, or the ***zeros*** of $\Delta(\lambda)$.

Theorem 1.1 : A continuous system S is asymptotic stable, if and only if, all the roots of $\Delta(\lambda) = 0$ are clustered in the open left half complex plane ($Re[\lambda_i] < 0$). A discrete system S is asymptotic stable, if and only if, all the roots of $\Delta(\lambda) = 0$ are clustered in the open unit disk ($|\lambda_i| < 1$).

The importance of Theorem 1.1 lies in the fact that asymptotic stability, which is a dynamic property, has an algebraic meaning, namely, clustering of the characteristic polynomial roots in certain regions in the complex plane. We call this property ***root-clustering*** . Since, in asymptotic stability, we are not concerned with the exact locations of the roots, it is reasonable to search for an algebraic test for root-clustering. The first algebraic test was presented by Hermite in 1854. We will review this and other classical results in Chapter 2.

1.3 ROOT-CLUSTERING AND PARAMETER SPACE

Asymptotic stability, although fundamental in dynamical systems, is not sufficient for good behavior. In a continuous system, if a root of $\Delta(\lambda)$ is too close to the imaginary axis ($Re[\lambda] = 0$), the damping ratio is small, and the system decays to its origin very slowly. In order to keep the roots far from the imaginary axis, we use the concept of ***relative stability.*** This concept generates the general problem of root-clustering, far beyond our needs in systems design. To be more specific, consider systems (1.1) - (1.3). With each ***free*** system we associate a matrix A or a characteristic polynomial $\Delta(\lambda)$. The collection of the eigenvalues λ_i of A is called the ***spectrum*** σ of A

$$\text{(i)} \quad \sigma(A) = \{\lambda \in \mathbb{C} : |\lambda I - A| = 0\}$$

$$\text{(ii)} \quad \sigma(\Delta) = \{\lambda \in \mathbb{C} : \Delta(\lambda) = 0\} . \tag{1.7}$$

Let $f(x,y) \in \mathbb{R}[x,y]$ be a real polynomial in the two variables x and y.We define an algebraic region $\aleph$ in the complex plane as one satisfying :

$$\text{(i)} \quad \aleph = \{(x+iy): f(x,y) < 0\}$$

$$\text{(ii)} \quad cl(\aleph) = \{(x+iy) : f(x,y) \le 0\} \tag{1.8}$$

$$\text{(iii)} \quad \partial\aleph = \{(x+iy): f(x,y) = 0\} .$$

The Root-Clustering Problem.

Given a matrix $A \in \mathbb{C}^{n \times n}$ and an algebraic region $\aleph$, find a test for

$$\sigma(A) \subset \aleph . \tag{1.9}$$

At this point we do not specify the nature of the test. However, as we shall see in chapter 3, there are two basic possibilities for tests. The first consists of a finite number of algebraic ***steps***, by means of which we can verify (1.9) for ***any*** A and $\aleph$. This approach, however, is unacceptable from a computational point of view. The second test consists of a finite number of algebraic (polynomial) ***inequalities*** (in the elements a_{ij} of A), like the Routh-Hurwitz criterion. In this case, we have to restrict the family of the allowed regions. We shall use the term ***criterion*** for such a test.

Two classical regions are the ***left half plane*** (LHP)

$$\aleph = \{(x+iy): x < 0\} \tag{1.10}$$

and the ***unit disk*** (UD)

$$\aleph = \{(x+iy): -1+x^2+y^2 < 0\} . \tag{1.11}$$

Well known tests for these two regions will be reviewed in Chapter 2.
A second important issue is the image of ***root-clustering in parameter space*** . If the matrix A is a function of some physical parameters, we define the space spanned by these parameters as parameter space. We then look for the region in this space for which (1.9) holds. In the process of designing a dynamical system, the designer may leave some parameters free. He then may choose values according to relative stability requirements. One important application is feedback design.

1.4 LINEAR FEEDBACK EXAMPLES

To open the discussion, consider a feedback system as shown in Figure 1.1. The plant has transfer function G(s), and the controller has a fixed structure $C(s) = k_1C_1(s)$ with a design parameter k_1. Two questions may be asked. First, what is the range of k_1 (if any) such that the closed loop has some prespecified relative stability propertis ? Second, what is the minimum value of k_1 such that these properties are met ?

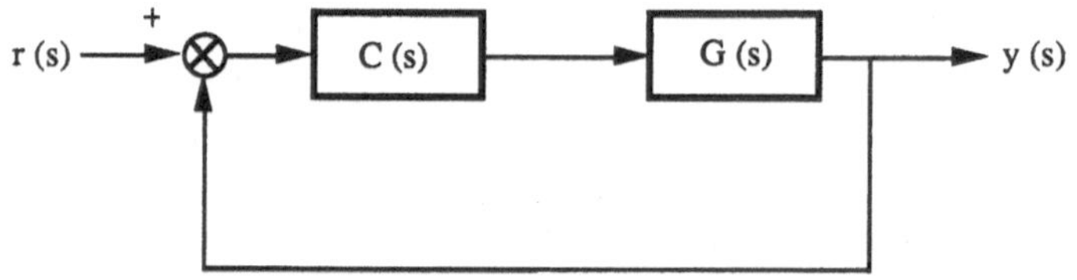

Figure 1.1 : A unit feedback system.

Relative stability can be measured, for example, using the root-clustering region presented in Figure 1.2. This region puts a lower bound on both the exponential decay rate and the damping ratio of the closed loop response. It is possible, of course, to replace the above region by a smooth one, like a hyperbola, as in Figure 1.3. Concerning the value of k_1, one may wish to minimize it in order to prevent saturation. On the other hand, the designer may minimize the integral square error :

$$I(k_1) = \int_0^\infty e^2(t)dt \tag{1.12}$$

to improve tracking. This may call for high gain values.

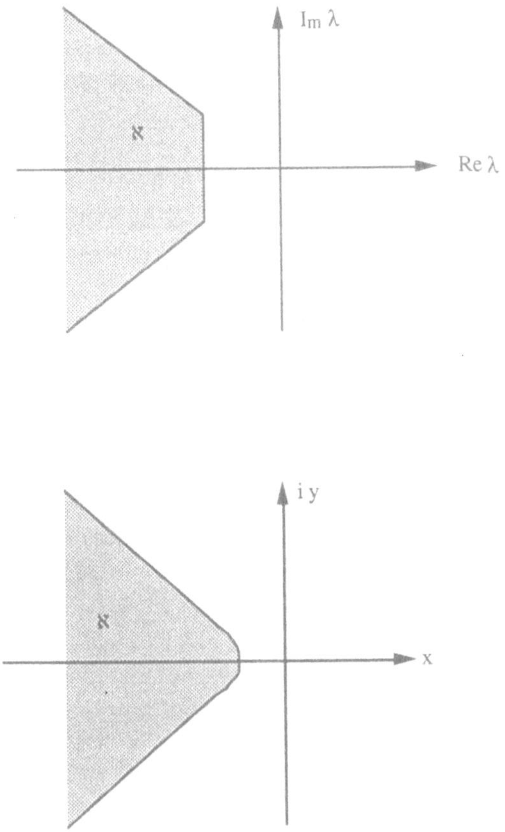

Fig. 1.3 : The left hyperbola.

Motivated by the above discussion, we propose

General Design Problem

Let $\Delta(\lambda)$ be a characteristic polynomial whose coefficients are polynomials in $k = [k_1,k_2,\ldots,k_m]'$. Solve the problem

$$\text{Min } \phi(k)$$
$$\text{s.t. } \sigma(\Delta) \subset \aleph \qquad (1.13)$$

for specified $\aleph$ and $\phi(\cdot)$.

Note that (1.13) is a multi-parameter problem in the gain space R^m and may include the design of the entire compensator C(s) in Figure 1.1. In Chapter 2, after reviewing root-clustering for the left half plane and the unit disk, we will review formulas for the integral square of signals. We will, then, update problem (1.13) and solve a numerical example. As a final note, the state space version of the above discussion is left to the reader as an exercise.

We next consider a problem in feedback control which to date does not have a closed form solution. Once more, consider the single input single output system given in Figure 1.1 . Let the plant and the compensator transfer functions be :

$$\text{(i)} \quad G(s) = \frac{n(s)}{d(s)}$$

$$\text{(ii)} \quad C(s) = \frac{x(s)}{y(s)} \tag{1.14}$$

respectively, where :

$$\text{(i)} \quad n(s) = \sum_{i=0}^{n} n_i s^i$$

$$\text{(ii)} \quad d(s) = \sum_{i=0}^{n} d_i s^i \; . \tag{1.15}$$

If n(s) and d(s) are relatively prime (i.e. do not have a common factor), then there exists a compensator C(s) of order n-1 such that the closed loop is stable. Moreover, we can find C(s) such that the closed loop characteristic polynomial is arbitrary. To see this, let :

$$\text{(i)} \quad x(s) = \sum_{i=0}^{n-1} x_i s^i$$

$$\text{(ii)} \quad y(s) = \sum_{i=0}^{n-1} y_i s^i \tag{1.16}$$

and let the required closed loop characteristic polynomial be

$$\Delta(s) = \sum_{i=0}^{2n-1} \Delta_i s^i \; . \tag{1.17}$$

The closed loop characteristic polinomial is given by :

$$n(s)x(s) + d(s)y(s) = \Delta(s) \; . \tag{1.18}$$

Equating coefficients of the same order of s in (1.18), we find :

$$[x_{n-1} x_{n-2}, \ldots x_0 \; y_{n-1} y_{n-2}, \ldots y_0] \; S(n, d) = [\Delta_{2n-1} \Delta_{2n-2}, \ldots \Delta_1 \Delta_0] \tag{1.19}$$

where S(n,d) takes the form

$$S(n,d) = \left[\begin{array}{cccccccc} n_n & n_{n-1} & n_{n-2} \cdots n_1 & n_0 & 0 & 0 \ldots 0 \\ 0 & n_n & n_{n-1} \cdots n_2 & n_1 & n_0 & 0 \ldots 0 \\ \vdots & & & & & \\ & & n_n \; n_{n-1} \; n_{n-2} & \cdots & n_1 & n_0 \\ \hline d_n & d_{n-1} & d_{n-2} \cdots d_1 & d_0 & 0 & 0 \ldots 0 \\ 0 & d_n & d_{n-1} \cdots d_2 & d_1 & d_0 & 0 \ldots 0 \\ \vdots & & & & & \\ & & d_n \; d_{n-1} \; d_{n-2} \cdots & & d_1 & d_0 \end{array}\right] . \tag{1.20}$$

It is known (Theorem 5.2) that S(n,d) is nonsingular, if and only if n(s) and d(s) are relatively prime. Thus, (1.19) has a unique solution for the compensators coefficients $\{x_i\}$ and $\{y_i\}$, if and only if n(s) and d(s) are relatively prime. The design of C(s) using (1.19) is called ***pole placement***. In some practical cases, however, a compensator of order n-1 is not acceptable, so we may ask the following two questions :

(i) What is the lowest order of stabilizing compensator ?

(ii) How can such a compensator be found ?

In some cases the lowest order is zero (proportional control), while in others, it is n-1, as above. Clearly, we have a wide range of possibilities from the lowest order to n-1. To solve this problem we first assume the compensator's order, say m, and write down the stability inequalities (Chapter 2) in terms of $\{x_i\}$ and $\{y_i\}$. Based on our results in Chapters 5 and 6, we will present, in Chapter 7, a solution to this multivariate set of inequalities, along with relative stability requirements. As a result, we give a definite answer to the above two questions.

1.5 AN AEROELASTIC EXAMPLE

In the investigation of aeroelastic stability, it is convenient to replace the original eigenvalue λ by a new varriable. In the new formulation it is required that the system's matrix A has all its eigenvalues clustered to the left of the ***stability parabola*** $y^2 = -\varepsilon^2 x$, where ε is a damping parameter. As an example, we consider a multilayered anisotropic rectangular flat panel exposed to a coplanar supersonic flow, at zero flow angle. In the linear theory, based on two modes, the system's matrix has the form :

$$A = \begin{bmatrix} \alpha_{11} & \alpha_{12}f & \alpha_{13}f & \alpha_{14} \\ \alpha_{21}f & \alpha_{22} & \alpha_{23} & \alpha_{24}f \\ \alpha_{31}f & \alpha_{32} & \alpha_{33} & \alpha_{34}f \\ \alpha_{41} & \alpha_{42}f & \alpha_{43}f & \alpha_{44} \end{bmatrix} \tag{1.21}$$

where f is the velocity parameter, and α_{ij} are functions of other parameters. The flutter speed limit f* is defined as the speed f for which the stability boundary of A is reached at two conjugate complex points. The design problem is two-fold :

(i) Find conditions on the parameters such that :

$$\sigma(A) \subset \{(x+iy): \varepsilon x+y^2 < 0\}, \tag{1.22}$$

where A is given by (1.21) ;

(ii) Choose the parameters such that f^* is maximized.

1.6 ABOUT THE BOOK

After presenting some basic concepts, we now give a brief outline of the book. In Chapter 2 we review basic structures for root clustering criteria with respect to the left half plane and the unit disk. These criteria are called ***classical*** since they deal with stability of continuous and discrete time linear systems. One measure of performance of linear systems is the integral square of signals. We use it in a minimization problem in the parameter (or gain) space. We finally demonstrate that an optimal response of a second order system is related to root clustering with respect to the left hyperbola. We open Chapter 3 by discussing the structure of root clustering criteria. We adopt the simplest structure, namely a single set of inequalities in the systems parameters. We then present criteria based on one-variable rational and irrational transformations. In chapter 4 we define the notion of transformability, based on two-variable transformations. We discuss three possibilities : polynomial, rational, and irrational transformability, and show that they all lead to root clustering criteria. We present a test for polynomial transformability and define an important class of regions belonging to rational transformability. We close this chapter with some non-trivial examples. In Chapter 5 we introduce composite matrices and composite polynomials, the former due to Stephanos, and the later developed while writing this book. We combine these results with those of Chapter 4 to obtain root clustering criteria for polynomial, rational, and irrational transformable regions. We present both matrix and polynomial versions. Next, in Chapter 6, we develop a criterion based on a matrix equation. This can be thought of as a generalization of the Lyapunov equation. For the polynomial version we replace the matrix equation by a modular algebra and save the computation effort. In many applications, systems data depends on physical parameters. The designer has to select these parameters in order to achieve some prespecified performances. A simple example is the compensator selection in linear feedback design. Chapter 7 is devoted to the important issue of parameter space. The key concept here is the so-called critical constraint. It is simply an equation which contains the image of the boundary in the parameter space. We describe an algorithm to isolate points in the parameter space belonging to the root clustering region. Using these results we construct fixed order compensators and present a robust analysis.

NOTES AND REFERENCES

In the last 40 years linear systems theory has gone a long way. Among many texts we mention Kailath [1], Chen [1], and Barnett [1]. The aeroelastic example of Section 1.5 is taken from Liberescu [1].

Chapter 2 : REVIEW OF CLASSICAL RESULTS

The use of the words ***classical results*** , here, means that we are reviewing asymptotic stability for (linear) continuous and discrete systems. In other words, given a polynomial $\Delta(\lambda)$, we present algebraic tests for the inclusion

$$\sigma(\Delta) \subset \aleph$$

for the left half plane

$$\aleph = \{(x+iy): x < 0\}$$

and for the unit disk

$$\aleph = \{(x+iy): -1+x^2 + y^2 < 0\}.$$

The form of the test is a set of ***polynomial inequalities*** in the coefficients of $\Delta(\lambda)$. For structural reasons, we will not present these results in historical order.

Consider the ***real*** polynomial

$$\Delta(\lambda) = \sum_{i=0}^{n} a_i \lambda^i \quad ; \quad a_n = 1 \ . \tag{2.1}$$

2.1 TABLE FORM

Given a polynomial (2.1), insert the rows

$$\begin{matrix} 1 & a_{n-2} & a_{n-4} & \cdots \\ a_{n-1} & a_{n-3} & a_{n-5} & \cdots \end{matrix}$$

as the first two rows in the following table :

$$\begin{matrix} s_{01} & s_{02} & s_{03} & \cdots \\ s_{11} & s_{12} & s_{13} & \cdots \\ s_{21} & s_{22} & \cdots & \\ s_{31} & \cdots & & \\ \vdots & & & \\ s_{n1} \ . & & & \end{matrix} \qquad s_{ij} \stackrel{\Delta}{=} s_{i-2,\,j+1} - \frac{s_{i-2,1}\, s_{i-1,\,j+1}}{s_{i-1,\,1}} \qquad i = 2, 3, \ldots, n \tag{2.2}$$

The following was proved by Routh in 1877.

Theorem 2.1 The roots of (2.1) are all clustered in the open left half complex plane, if and only if, all the elements of the first column $\{s_{i1}\}$ in the Routh Table (2.2) are positive ($s_{i1} > 0$, $i = 0,1,\ldots,n$).

For many years researchers have tried to develop a Routh-type table test for root-clustering in the unit circle. The following version was developed in 1985 by Chen and Chan.

Given a polynomial (2.1), insert the rows

$$\begin{array}{llll} a_0 & a_1 & \cdots\; a_{n-1} & 1 \\ 1 & a_{n-1} & \cdots\; a_1 & a_0 \end{array}$$

as the first two rows in the following table :

$$\begin{array}{llllll} s_{01} & s_{02} & s_{03} & \cdots & s_{0n} & s_{0,n+1} \\ t_{01} & t_{02} & t_{03} & \cdots & t_{0n} & t_{0,n+1} \\ s_{11} & s_{12} & s_{13} & \cdots & s_{1n} & \\ t_{11} & t_{12} & t_{13} & \cdots & s_{1n} & \\ \vdots & & & & & \\ s_{n1} & & & & & \\ t_{n1} & & & & & \end{array} \qquad (2.3)$$

where

(i) any t row is the reversed (order) s row,

$$\text{(ii)} \quad s_{ij} = s_{i-1,j+1} - \frac{t_{i-1,j+1}\, s_{i-1,1}}{t_{i-1,1}} \qquad \begin{array}{l} i = 1,2,\ldots,n \\ \\ j = 1,2,\ldots,(n-i+1)\,. \end{array} \qquad (2.4)$$

Theorem 2.2 The roots of (2.1) are all clustered in the open unit disk, if and only if, all the elements of the first column of the t rows $\{t_{i1}\}$ in table (2.3) are positive ($t_{i1} > 0$, $i = 0,1,\ldots,n$).

2.2 HERMITE FORM

Given a polynomial (2.1), we define the nxn Hermite matrix $H = [h_{ij}]$, defined by :

$$\begin{aligned} h_{ij} &= \sum_{k=1}^{i} (-1)^{k+i} a_{n-k+1} a_{n-i-j+k} && j \geq i\,,\; i+1 = \text{even} \\ &= h_{ji} && j > i\,,\; j+i = \text{even} \qquad (2.5) \\ &= 0 && j+i = \text{odd}\,. \end{aligned}$$

As early as 1854, Hermite obtained the following result.

Theorem 2.3 The roots of (2.1) are all clustered in the open left half complex plane, if and only if, the Hermitian (in fact, symmetric) matrix H defined in (2.5) is positive definite.

We now proceed to the unit circle. With the real polynomial (2.1), we associate the nxn symmetric matric $C = [c_{ij}]$, defined by :

$$c_{ij} = \sum_{p=1}^{\min(i,\, j)} (a_{n-i+p} a_{n-j+p} - a_{i-p} a_{j-p} \, . \tag{2.6}$$

Then, following Schur and Cohn we have

Theorem 2.4 The roots of (2.1) are all clustered in the open unit disk, if and only if, the symmetric matrix C defined in (2.6) is positive definite.

2.3 HURWITZ FORM

The Hermitian forms (2.5) and (2.6) are second order in the polynomial's coefficients $\{a_i\}$. In 1895, Hurwitz obtained a root-clustering criterion for the left half plane which is first order in the a_i's . With the polynomial (2.1) we associate the nxn Hurwitz matrix H, defined by :

$$H = \begin{bmatrix} a_{n-1} & a_{n-3} & a_{n-5} & \cdots & 0 \\ 1 & a_{n-2} & a_{n-4} & \cdots & 0 \\ 0 & a_{n-1} & a_{n-3} & \cdots & 0 \\ 0 & 1 & a_{n-2} & \cdots & 0 \\ \vdots & \vdots & \vdots & & \vdots \\ & & & \cdots & a_0 \end{bmatrix} . \tag{2.7}$$

Theorem 2.5 The roots of (2.1) are all clustered in the open left half complex plane, if and only if, in the matrix H defined in (2.7), all the principle minors are positive.

2.4 INNER FORM

It is known that necessary conditions for root-clustering in the left half plane are $a_i > 0$, $i = 0,1,\ldots,n-1$. If we use this fact, we do not need all the principle minors in the previous criteria. This was first observed by Leonard and Chipart. A natural form, in which the number of subdeterminants needed is minimal, is the Inner form due to Jury. For continuous-time systems, we associate with (2.1) the matrix :

$$\Delta^e_{n-1} = \begin{bmatrix} a_n & a_{n-2} & a_{n-4} & \cdots & \Delta_1 & \Delta_3 \\ 0 & a_n & a_{n-2} & \cdots & & \\ & & a_n & a_{n-2} & a_{n-4} & \cdots \\ & \bigcirc & 0 & a_{n-1} & a_{n-3} & \cdots \\ & & a_{n-1} & a_{n-3} & a_{n-5} & \cdots \\ 0 & a_{n-1} & a_{n-3} & \cdots & & \\ a_{n-1} & a_{n-3} & a_{n-5} & \cdots & & \end{bmatrix} \tag{2.8}$$

for n even, and the matrix

$$\Delta^{0}_{n-1} = \begin{bmatrix} a_{n-1} & a_{n-3} & a_{n-5} & \cdots & \Delta_2 \\ 0 & a_{n-1} & a_{n-3} & \cdots & \\ & \bigcirc & a_{n-1} & a_{n-3} & \cdots \\ & & a_n & a_{n-2} & \cdots \\ 0 & a_n & a_{n-2} & \cdots & \\ a_n & a_{n-2} & a_{n-4} & \cdots & \end{bmatrix} \tag{2.9}$$

for n odd. We then have

Theorem 2.6 The roots of (2.1) are all clustered in the open left half complex plane, if and only if, $a_i > 0$, $i = 0,1,\ldots,n-1$, and

(i) Δ^{e}_{n-1} is positive innerwise: $|\Delta_1| > 0$, $|\Delta_3| > 0, \ldots, |\Delta_{n-1}| > 0$, for n even;
(ii) Δ^{0}_{n-1} is positive innerwise: $|\Delta_2| > 0$, $|\Delta_4| > 0, \ldots, |\Delta_{n-1}| > 0$, for n odd.

It should be noted that we may require only half of the a_i's to be positive (in proper order); however, the condition $a_i > 0$ is simple — it can be verified by inspection.

For the discrete-time system, we associate with (2.1) the matrices

$$\Delta^{\pm}_{n-1} = X_{n-1} \pm Y_{n-1} \tag{2.10}$$

where

$$X_{n-1} = \begin{bmatrix} a_n & a_{n-1} & a_{n-2} & \cdots & a_2 \\ & a_n & a_{n-1} & \cdots & a_3 \\ & \bigcirc & & \ddots & \vdots \\ & & & & a_n \end{bmatrix} \qquad Y_{n-1} = \begin{bmatrix} & & & & a_0 \\ & \bigcirc & & \cdot^{\cdot^{\cdot}} & \vdots \\ & a_0 & a_1 & \cdots & a_{n-1} \\ a_0 & a_1 & a_2 & \cdots & a_{n-2} \end{bmatrix} . \tag{2.11}$$

The following was established by Jury in 1970.

Theorem 2.7 The roots of (2.1) are all clustered in the open unit disk, if and only if,

(i) $\Delta(1) > 0$, $(-1)^n \Delta(-1) > 0$;

(ii) $\Delta^{\pm}_{n-1}$ are positive innerwise.

2.5 LYAPUNOV FORM

The previous forms were constructed for polynomials. In case we wish to test the eigenvalues of a matrix using these forms, we have to calculate the characteristic polynomial first. In 1892, as a special case of a nonlinear theory, Lyapunov obtained a ***direct*** result.

Theorem 2.8 The eigenvalues of a square complex matrix A (the roots of (1.6ii) are all clustered in the open left half plane, if and only if, given any positive definite (p.d.) matrix $Q=Q^*$, the unique solution $P = P^*$ of

$$PA + A^*P = -Q \tag{2.12}$$

is p.d.

Unlike the previous tests, this test has an identical structure for both real and complex matrices. Here, Theorem 2.8 is stated for a complex matrix A. In 1952, Stein presented the discrete counterpart.

Theorem 2.9 The eigenvalues of a square complex matrix A are all clustered in the open unit disk, if and only if, given any positive definite matrix $Q = Q^*$, the unique solution $P = P^*$, satisfying

$$A^*PA - P = -Q \tag{2.13}$$

is p.d.

Theorems 2.8 and 2.9 are stability tests with respect to the system's matrix A, without the need to calculate the characteristic polynomial. Since in both (2.12) and (2.13), the left hand side is linear in P, both equations are equivalent to a set of linear equations in P_{ij}. In fact, if we stack the rows of P and Q into two column vectors p and q, the above equations are equivalent to :

$$\phi(A)p = -q \tag{2.14}$$

where

$$\phi(A) = A \otimes I + I \otimes \bar{A} \tag{2.15}$$

for (2.12), and

$$\phi(A) = A \otimes \bar{A} - I \tag{2.16}$$

for (2.13). The symbol $\otimes$ denotes the Kronecker product, and is defined by :

$$A \otimes B = [a_{ij}B] = \begin{bmatrix} a_{11}B & a_{12}B & \cdots \\ a_{21}B & & \\ \vdots & & \end{bmatrix} \tag{2.17}$$

Remark: The eigenvalues of $\phi(A)$ in (2.15) are $\{\lambda_i + \bar{\lambda}_j\}$; i, j = 1, 2,..., n. The eigenvalues of $\phi(A)$ in (2.16) are $\{\lambda_i\bar{\lambda}_j - 1\}$; i, j = 1, 2,..., n. The proof for this result will be given in Section 5.2 for a more general case.

2.6 SOME PROOFS

In this section we first sketch the proofs of Theorems 2.8, 2.9 and then present the connections to the rest of the tests.

Proof of Theorem 2.8

Sufficiency : Let z_i be the eigenvector of A corresponding to the eigenvalue λ_i. Multiplying (2.12) from left and right by z_i^* and z_i, respectively, and using $Az_i = \lambda_i z_i$, $z_i^*A^* = \lambda_i z_i^*$, we obtain :

$$(\lambda_i + \bar{\lambda}_i)z_i^*Pz_i = -z_i^*Qz_i\,. \tag{2.18}$$

Thus, if Q > 0 (p.d.) implies P > 0, it follows that

$$\lambda_i + \bar{\lambda}_i = 2\,\mathrm{Re}\,\lambda_i < 0 \qquad \forall i\,. \tag{2.19}$$

Necessity : This part consists of three main steps :

Step 1. Uniqueness of solution: By hypothesis $\mathrm{Re}(\lambda_i) < 0$ $\forall$ i. Thus, $\lambda_i + \bar{\lambda}_j \neq 0$ $\forall$ i and j. On the other hand, the eigenvalues of $\phi(A)$ in (2.15) are $\{\lambda_i + \bar{\lambda}_j\}$. Thus, $\phi(A)$ is nonsingular. This implies that (2.14) and (2.12) have unique solutions.

Step 2. There exists a p.d. pair $\{P_o, Q_o\}$ satisfying (2.12): Since the eigenvalues of A are invariant under similarity transformation, we can write (2.12) with respect to J, the Jordan canonical form. Indeed, if $J = T^{-1}AT$, substitute $A = TJT^{-1}$ in (2.12). Multiplying (2.12) from the left by T* and from the right by T, we obtain $FJ+J^*F = -K$, where $F = T^*PT$, $K = T^*QT$. Next, we use a second similarity transformation $G = D^{-1}JD$, with $D = \mathrm{diag}[1\ \delta\ \delta^2 \ldots \delta^{n-1}]$. As a result we have $G = \Lambda + \delta U$, where $\Lambda = \mathrm{diag}[\lambda_1\ \lambda_2, \ldots, \lambda_n]$, and the elements of U are either 0 or 1, with all nonzero elements located on the diagonal above the main diagonal.

Using $P = P_o = \mathrm{diag}[-(\lambda_1 + \bar{\lambda}_1), -(\lambda_2 + \bar{\lambda}_2), \ldots, -(\lambda_n + \bar{\lambda}_n)]$ in $P_oG + G^*P_o = -Q$ we find $P_oG + G^*P_o = -\mathrm{diag}[(\lambda_1 + \bar{\lambda}_1)^2, -(\lambda_2 + \bar{\lambda}_2)^2, \ldots, -(\lambda_n + \bar{\lambda}_n)^2] + 0(\delta)$. By hypothesis, $2\mathrm{Re}(\lambda_i) = \lambda_i + \bar{\lambda}_i < 0$. Thus P_o is p. d.. By continuity, for small δ, $P_oG + G^*P_o$ is negative definite. Thus, there exist p.d. matrices P_o, Q_o satisfying (2.12).

Step 3. Arbitrary p.d. Q: For an arbitrary p.d. Q, let $Q_t = tQ + (1-t)Q_o$, with $0 \le t \le 1$. Clearly Q_t is p.d.. For $Q = Q_t$, equation (2.12) has a unique solution P_t which is continuous in t. It is left to prove that P_t never becomes singular. If P_t does become singular there exists an orthogonal transformation

$$F_t = T^*P_tT = \begin{bmatrix} P_t & 0 \\ 0 & 0 \end{bmatrix}$$

Equation (2.12) has the form $F_tH + H^*F_t = -K$. The nxn element of the left hand side is zero, while that of the right hand side is negative. This contradiction implies that P_t never becomes singular, and the proof is completed.

Proof of Theorem 2.9

Sufficiency: as above, but (2.18) is replaced by :

$$(\lambda_i\bar{\lambda}_i - 1)z_i^*Pz_i = -z_i^*Qz_i . \tag{2.20}$$

Thus, $Q>0$ and $P>0$ imply :

$$\lambda_i\bar{\lambda}_i - 1) = |\lambda_i|^2 - 1 < 0, \text{ or}$$

$$|\lambda_i|^2 < 1 \quad \forall i . \tag{2.21}$$

Necessity: as above, but $P_0 = \mathrm{diag}[(1-|\lambda_1|^2)\,(1-|\lambda_2|^2)\,...(1-|\lambda_n|^2\,]$, and

$$G^*P_0G - P_0 = -\mathrm{diag}\,[(1-|\lambda_1|^2)\,(1-|\lambda_2|^2)\,...(1-|\lambda_n|^2\,] .$$

The reader who is familiar with other proofs of Theorem 2.8 may object to the length of the present proof for the necessity part. Indeed, for that part, we may write the solution of (2.12) as

$$P = \int_0^\infty e^{A^*t}Qe^{At}\,dt . \tag{2.22}$$

Since, by hypothesis $\sigma(A) \subset$ LHP, the integral exists. Thus, given any Hermitian p.d. matrix Q, P is also p.d. . Similarly, the solution of (2.13) is :

$$P = \sum_0^\infty A^{*i}QA^i . \tag{2.23}$$

A more complicated proof is needed for the general case of root-clustering. In this case, we do not have a solution for P similar to (2.22). Thus, it is constructive to use an algebraic proof at this preliminary stage.

Proofs of other tests

While Theorems 2.1 - 2.7 are related to the characteristic polynomial, Theorems 2.8 - 2.9 are related to the matrix A. Instead of constructing a direct proof for the polynomial version as was originally done by Hermite, Routh and others, we will show the connection between the Lyapunov and the Table forms. The reader should bear in mind, however, that the Lyapunov form for matrices implies the polynomial version in a straightforward way. Given the ***real*** polynomial (2.1), form the ***companion*** matrix

$$A = \begin{bmatrix} 0 & 1 & 0 & \dots & 0 \\ 0 & 0 & 1 & \dots & 0 \\ \vdots & \vdots & \vdots & & \vdots \\ 0 & 0 & 0 & & 1 \\ -a_0 & -a_1 & -a_2 & \dots & -a_{n-1} \end{bmatrix}$$

(2.24)

Since the characteristic polynomial of A is (2.1), we may apply directly Theorems 2.8 or 2.9. In this way we should find tests, equivalent to Theorems 2.1 - 2.7. However, what we need is a direct test on the coefficients $\{a_i\}$, like the Table form. To demonstrate the validity of the Table Form, we need a sufficient condition which is weaker than Theorems 2.8, 2.9. Returning to equation (2.18) we see that if P is p.d., and $z^*_i Q z_i > 0$ $\forall i$, then $\sigma(A) \subset$ LHP. However, a direct substitution shows that $Az = \lambda z$ is satisfied with A in companion form (2.24), and $z = [1 \ \ \lambda \ \ \lambda^2 \dots \lambda^{n-1}]'$. Thus, if (2.12) is satisfied with A given by (2.24), $Q = ee'$, $e' = [1 \ \ 0 \dots 0]$, and a p.d. P, then the roots of (2.1) are all clustered in the open LHP. The converse is also true. If $\sigma(\Delta) \subset$ LHP, then $\sigma(A) \subset$ LHP, with A given by (2.24). Thus, the integral in (2.22) exists. Using the above Q in (2.22) results in a p.s.d. P ($P \geq 0$), and possibly p.d.. It is left to prove that P is not singular. Similar to the proof of Theorem 2.8, if P is singular we can transform it to $F = T'PT$, with orthogonal T. Then (2.12) is equivalent to $FH + H'F = -K$, where $H = T'AT$, $K = T'QT$. If we choose T such that

$$F = \begin{bmatrix} \tilde{P}_1 & 0 \\ 0 & 0 \end{bmatrix}$$

we once more find that the nxn element of the left hand side is zero, while that of the right hand side is negative.This contradiction implies that P is nonsingular and in fact p.d.. The reader may verify similar results for the discrete case. To summarize our results, we state the following theorem.

Theorem 2.10 The roots of the real polynomial (2.1), are all clustered in the open left half plane, if and only if, for $Q = ee'$, $e' = [1 \ \ 0 \dots 0]$ and the companion matrix (2.24) associatedwith (2.1), the unique solution P of $PA + A'P = -Q$ is p.d.. A similar result holds for the unit disk.

The reader can verify that $PA + A'P = -Q$ is equivalent to $FH + H'F = -K$, with $H = T^{-1}AT$, $F = T'PT$ and $K = T'QT = T'ee'T \stackrel{\Delta}{=} aa'$. Since for A in companion form (2.24) and $e' = [1 \ \ 0 \dots 0]$ the pair $\{A',e\}$ is controllable, and since controllability and spectrum are invariant under similarity transformation, we can generalize Theorem 2.10 to obtain:

Theorem 2.11 The eigenvalues of a real matrix A are all clustered in the open left half plane, if and only if, given any p.s.d. $Q = ee'$ with $\{A',e\}$ controllable, the unique solution P of $PA + A'P = -Q$ is p.d.. A similar result holds for the unit disk.

To prove Theorem 2.1 we use the Schwarz form:

$$S = \begin{bmatrix} 0 & 1 & 0 & \dots & 0 \\ -b_0 & 0 & 1 & \dots & 0 \\ 0 & -b_1 & 0 & \dots & 1 \\ \vdots & \vdots & \vdots & & \vdots \\ 0 \dots & 0 & -b_{n-2} & & -b_{n-1} \end{bmatrix} \tag{2.25}$$

and choose:

$$e' = \left[\, 0 \dots 0 \;\; b_{n-1}\sqrt{2} \,\right] . \tag{2.26}$$

It is known that the Schwarz matrix is similar to the companion matrix and thus has the same eigenvalues. In particular, $S = T^{-1}AT$ where A and S are given by (2.24) and (2.25) respectively, and T is formed of the Routh entries s_{ij} by :

$$T^{-1} = \begin{bmatrix} 1 & & & & & \\ 0 & 1 & & & \bigcirc & \\ \frac{s_{n-2,2}}{s_{n-2,1}} & 0 & 1 & & & \\ 0 & \frac{s_{32}}{s_{31}} & 0 & 1 & & \\ \frac{s_{n-4,3}}{s_{n-4,1}} & 0 & \frac{s_{22}}{s_{21}} & 0 & 1 & \\ \vdots & \cdots \frac{s_{13}}{s_{11}} & 0 & \frac{s_{12}}{s_{11}} & 0 & 1 \end{bmatrix} . \tag{2.27}$$

The solution to the Lyapunov equation is

$$P = \text{diag}\left[\prod_{0}^{n-1} b_i \quad \prod_{1}^{n-1} b_i \dots b_{n-1} \right] . \tag{2.28}$$

We see that P is p.d. if and only if $b_i > 0 \quad \forall i$. Since

$$b_{n-1} = |H_1| , \; b_{n-2} = \frac{|H_2|}{|H_1|} , \; b_{n-3} = \frac{|H_3|}{|H_2||H_1|} , \dots b_{n-i} = \frac{|H_{i-3}||H_i|}{|H_{i-2}||H_{i-1}|} \tag{2.29}$$

where H_i are the principle minors of the Hurwitz matrix (2.7), Theorem 2.5 follows. The well known equivalence between the Routh and Hurwitz theorems implies Theorem 2.1. To prove Theorem 2.2 we transform the companion matrix using similarity transformation T, where :

$$T^{-1} = \begin{bmatrix} 1 & & & & \\ \dfrac{t_{n-1,2}}{t_{n-1,1}} & 1 & & \bigcirc & \\ \vdots & & & & \\ \dfrac{t_{2,n-1}}{t_{21}} & \cdots & \dfrac{t_{22}}{t_{21}} & 1 & \\ \dfrac{t_{1n}}{t_{11}} & \dfrac{t_{1,n-1}}{t_{11}} & \cdots & \dfrac{t_{12}}{t_{11}} & 1 \end{bmatrix} . \tag{2.30}$$

The resulting A matrix is called the discrete Schwarz matrix. If we use :

$$e' = \begin{bmatrix} 0 \ldots 0 & \dfrac{t_{11}t_{01}}{t_{0,n+1}} \end{bmatrix} \tag{2.31}$$

the solution to the Lyapunov equation becomes :

$$P = \operatorname{diag}\left[t_{n1} \quad t_{n-1,1} \cdots t_{21} \quad t_{11}\left(\frac{t_{01}}{t_{0,n+1}}\right)^2 \right] \tag{2.32}$$

Thus, P is p.d., if and only if t_{i1} are positive, as in Theorem 2.2. The Inners form is closely related to the Table form and is proved in a similar way.

To close this section we prove the Hermite form. For this case we choose the pair {A',e}, where A is the companion matrix (2.24), and

$$e' = \sqrt{2}\,[a_{n-1} \quad 0 \quad a_{n-3} \quad 0 \ldots] . \tag{2.33}$$

The use of the above pair in the Lyapunov equation results in P identical to H in (2.5).

2.7 INTEGRAL SQUARE OF SIGNALS

One way to measure the quality of a dynamical system is to evaluate the integral square of a certain signal, like the output of a free system or the error between a command input and the output. In case the designer has some free parameters, he may wish to minimize the integral with respect to these parameters.

Transfer Function Approach

Let g(t) be a continuous signal whose integral square is to be evaluated. Let G(s) = L[g(t)] be its Laplace Transform. According to Parseval,

$$I_n = \int_0^{\infty} g^2(t)dt = \frac{1}{2\pi j} \int_{-j\infty}^{j\infty} G(s)\, G(-s)ds \tag{2.34}$$

provided the integral exists. Define :

$$G(s)\, G(-s) = \frac{g(s)}{h(s)h(-s)} \tag{2.35}$$

where :

$$g(s) = b_{n-1}s^{2(n-1)} + b_{n-2}s^{2(n-2)} + \ldots + b_0 \tag{2.36}$$

$$h(s) = a_n s^n + a_{n-1}s^{n-1} + \ldots + a_0 . \tag{2.37}$$

Note that g(s) is an even polynomial and h(s) is the characteristic polynomial. It is known that I_n is given by :

$$I_n = \frac{(-1)^{n+1}}{2a_n} \cdot \frac{|W_n|}{|H_n|} \tag{2.38}$$

where H_n is the Hurwitz matrix (2.7), and W_n is generated from H_n by replacing the ***first row*** by $[b_{n-1} \ldots b_0]$.

For the discrete case, let $\{f_k\}$ be a sequence with f(z) as the z-transform. The Parseval formula takes the form :

$$I_n = \sum_{k=0}^{\infty} f_k^2 = \frac{1}{2\pi j} \oint_{\substack{\text{unit}\\ \text{disk}}} f(z)\, f(z^{-1}) \frac{dz}{z} \tag{2.39}$$

provided the integral exists. Define:

$$f(z) = \frac{\sum_{i=1}^{n} b_1 z^i}{\sum_{i=0}^{n} a_i z^i} , \tag{2.40}$$

then :

$$I_n = \frac{1}{a_n} \frac{|\Delta_{n+1}^{+}|_b}{|\Delta_{n+1}^{+}|} \tag{2.41}$$

where $\Delta_{n+1}^{+} \;=\; X_{n+1} + Y_{n+1}$,

$$X_{n+1} = \begin{bmatrix} a_n & a_{n-1} \cdots a_0 \\ & a_n \quad \cdots \quad a_1 \\ \bigcirc & \vdots \\ & a_n \end{bmatrix} \qquad Y_{n+1} = \begin{bmatrix} \bigcirc & & a_0 \\ & & \vdots \\ & a_0 \cdots & a_{n-1} \\ a_0 & a_1 \cdots & a_n \end{bmatrix}$$

(2.42)

and where $|\Delta^+_{n+1}|_b$ is generated from $|\Delta^+_{n+1}|$ by replacing the ***last row*** by :

$$[2b_n b_0 \quad 2\sum b_i b_{i+n-1} \cdots 2\sum b_i b_{i+1} \quad 2\sum_{i=0}^{n} b_i^2] .$$

The connection of the above integrals to stability is clear. Each integral exists if and only if the respective system is asymptotic stable. Observing the denominators of (2.38) and (2.41) we see that each integral tends to'infinity as we approach the boundary of stability.

State Space Approach

Consider the continuous free system :

$$\dot{x} = Ax, \qquad x(0) = x_0 . \tag{2.43}$$

We associate with this system the integral

$$I_n = \int_0^\infty x'(t)\, Qx(t)\, dt \tag{2.44}$$

where $Q = Q'$ is p.s.d., and possibly p.d. .

In order to evaluate the ingegral, we define the quadratic form :

$$V(x) = x'Px , \quad P = P'. \tag{2.45}$$

Then,

$$W(x) \triangleq \text{grad}_x V(x) \cdot \dot{x} = 2x'\, PAx = x'(PA + A'\, P)x . \tag{2.46}$$

Choose P such that :

$$PA + A'P = -Q \tag{2.47}$$

to obtain :

$$W(x) = -x'Qx . \tag{2.48}$$

A direct integration yields :

$$\int_0^\infty x'(t)Qx(t)dt = V(x_0) - V(\infty) = x'_0Px_0 - \lim_{T\to\infty} x'(T)Px(T) \quad .$$

If the system is asymptotically stable, then $x(t) \to 0$, and

$$I_n = x'_0Px_0 \, . \tag{2.49}$$

Next, consider the discrete free system

$$x_{k+1} = Ax_k \, . \tag{2.50}$$

We are interested in the sum :

$$I_n = \sum_{k=0}^{\infty} x'_k Qx_k \, . \tag{2.51}$$

As before, we define the quadratic form :

$$V_k = x'_k Px_k \, . \tag{2.52}$$

Then,

$$V_{k+1} = x'_{k+1} Px_{k+1} = x'_k A' PAx_k \, .$$

Construct the difference :

$$\Delta V \stackrel{\Delta}{=} V_{k+1} - V_k = x'_k(A' PA - P)x_k \, . \tag{2.53}$$

Choose P such that :

$$A'PA\text{-}P = \text{-}Q \tag{2.54}$$

to obtain :

$$\Delta V = - x'_k Qx_k \, . \tag{2.55}$$

Next, we evaluate the infinite sum on both sides of equation (2.55). The right hand side yields - I_n. The left side yields :

$$\lim_{m\to\infty} \sum_{k=0}^{m} \Delta V = \lim_{m\to\infty} \sum_{k=0}^{m} (V_{k+1} - V_k) = (V_1 - V_0) + (V_2 - V_1) + (V_3 - V_2) + \ldots$$

$$= - V_0 + \lim_{m\to\infty} V_{m+1} \, .$$

If the system is asymptotically stable, $x_k \to 0$, $\lim_{m \to \infty} V_{m+1} = 0$, and we are left with

$$I_n = x'_0 P x_0 . \tag{2.56}$$

2.8 STABILITY IN PARAMETER SPACE

In the introduction we met the root-clustering inclusion (1.9) and the minimization (1.13) in the parameter space. In light of the stability criteria presented previously, we wish to present more concrete expressions. We will limit our discussion to the left half plane and the unit disk. Later in the book we will generalize our results. One important fact becomes clear. Root-clustering in the left half plane and the unit disk consists of **a set of polynomial inequalities** with respect to the coefficients. Denote these inequalities by $\{\phi_k\}$. Then :

$$\begin{aligned} &\text{(i)} \quad \sigma(A) \subset \aleph \Leftrightarrow \{\phi_L(a_{ij}) > 0\}_{L=1}^{n} \\ &\text{(ii)} \quad \sigma(\Delta) \subset \aleph \Leftrightarrow \{\phi_L(a_i) > 0\}_{L=1}^{n} \end{aligned} \tag{2.57}$$

where $\aleph$ is either the LHP or the UD.

Now suppose that the coefficients of A (or Δ) are polynomial functions of certain parameters $\{k_i\}$, $i = 1,2,...,m$. The k's may be physical parameters in some applications and feedback gains in others. According to the specific application, we may wish to minimize some scalar function $\phi(k)$. We list a few examples :

1. $\phi(k) = \sum_{i=1}^{m} w_i^2 k_i^2$, w_i are given constants; (2.58)

2. $\phi(k) = I_n(k)$, given by (2.38) or (2.41); (2.59)

3. $\phi(k) = \text{tr}[P(k)]$, given by (2.47) or (2.54). (2.60)

The design problem (1.13) becomes :

$$\left\| \begin{array}{l} \text{Min } \phi(k) \\ \text{s. t. } \phi_i(k) \geq 0 , \qquad i = 1, 2, \ldots, n . \end{array} \right. \tag{2.61}$$

Note the change in the stability inequalities. Originally, these inequalities are strict, $\phi_i > 0$. This defines an ***open set*** in the parameter space R^m. In order to be able to use (2.58) , we have to use a closed set, so we use $\phi_i(k) \geq 0$. In most practical cases, the minimum for (2.58) is obtained on the stability boundary. But this should not discourage us. Later in this book we will develop relative stability criteria. In those cases, $\phi_i(k) \geq 0$ is certainly accepted. The situation with (2.59) and (2.60) is simpler, since the minimum is attained in the ***open set***

$$\tilde{\aleph} \overset{\Delta}{=} \{k \in R^m : \phi_i(k) > 0\ ,\ \ i = 1,2,\ldots,n\}\ . \tag{2.62}$$

$\tilde{\aleph}$ is the image of the stability region $\aleph$ in the parameter space. This region will be an important subject later in this book.

Example 2.1

Consider the feedback system of figure 2.1. It is desired to select the damping ratio ζ such that $\int_0^\infty y^2(t)dt$ is minimized

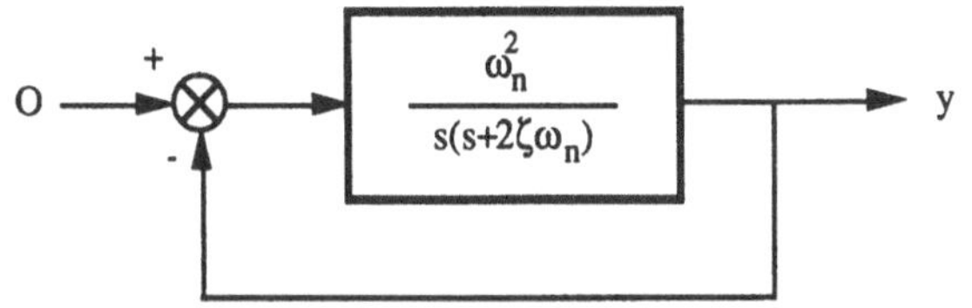

Figure 2.1

The state-space model of the closed loop system is :

$$\dot{x} = \begin{bmatrix} 0 & 1 \\ -\omega_n^2 & -2\zeta\omega_n \end{bmatrix} x$$

$$y = [1 \quad 0]\,x\ . \tag{2.63}$$

Using (2.44) , (2.47) and (2.49) we find for $Q = ee'$, $e' = [1 \quad 0]$,

$$P = \frac{1}{\omega_n}\begin{bmatrix} \zeta + \frac{1}{4\zeta} & \frac{1}{2\omega_n} \\ \frac{1}{2\omega_n} & \frac{1}{4\zeta\omega_n^2} \end{bmatrix}. \tag{2.64}$$

For (2.60) ,

$$\phi(\zeta) = \frac{1}{\omega_n}\left[\zeta + \frac{1}{4\zeta}\left(1 + \frac{1}{\omega_n^2}\right)\right]. \tag{2.65}$$

The minimum of $\phi(.)$ satisfies $\phi'(\zeta) = 0$. Thus,

$$\zeta = \sqrt{\frac{1}{4}\left(1 + \frac{1}{\omega_n^2}\right)}\ . \tag{2.66}$$

For asymptotic stability we require

$$\zeta\omega_n > 0 . \tag{2.67}$$

For the second order system (2.63), it is known that if

$$s = x + iy , \tag{2.68}$$

then

$$x = -\zeta\omega_n , \quad y = \omega_n\sqrt{1-\zeta^2} . \tag{2.69}$$

Simple calculations show that (2.66) implies

$$3x^2 - y^2 = 1 . \tag{2.70}$$

This defines an hyperbola in the complex plane. For asymptotic stability (2.67), we take the left branch, as shown in Figure 2.2.

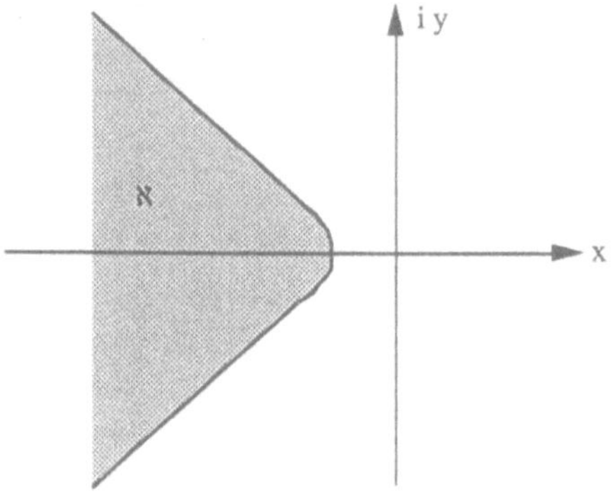

Figure 2.2 The left hyperbola

As the reader will have observed, this example is carefully constructed. For a second order system, the integral square of the output is minimized with respect to the damping ratio, if the eigenvalues lie on the hyperbola (2.70). Motivated by our results, we may require for higher order systems that the eigenvalues of the system's matrix A are all clustered to the left of the hyperbola (2.70), the shaded region in Figure 2.2 . This takes the general form :

$$\sigma(A) \subset \{(x + iy) : 1 - a^2x^2 + b^2y^2 < 0\} \tag{2.71}$$

which agrees with Figure 1.3.

NOTES AND REFERENCES

Many results can be found in Jury [1] and Barnett [1]. The table form of Theorem 2.1 is in Routh [1]. The discrete counterpart, Theorem 2.2, is taken from Chen and Chan [1]. See also Astrom [1], Marden [1], Jury [2], and Bistritz [1]. Hermite was a pioneer. In addition to a criterion for the half plane, his theory Hermite [1] implies root clustering with respect to regions which in later chapters are called "rank2 signature 0". An explicit criterion for the unit circle, Theorem 2.4, was established by Schur [1] and Cohn [1]. While the Hermitian matrices in Theorems 2.3 and 2.4 are second order in the a_i's, the matrix H in Theorem 2.5 due to Hurwitz [1] is first order in these coefficients. According to Lienard and Chipart [1], about one half of the principle minors of H can be replaced by the same number of coefficients a_i. This is presented here in Inner form for both continuous and discrete time systems, Jury [3]. The next form, Theorem 2.8, due to Lyapunov [1], is indeed interesting. Since the matrix Q in (2.12) is arbitrary positive definite, it gives rise to many possible structures. The discrete counterpart is due to Stein [1]. The solution of matrix equations (2.12) and (2.13) is a special case of a general theory of matrix equations, Macduffee [1]. The proofs of Theorems 2.8 and 2.9 are due to Jury and Ahn [1] and are based on some elements of Howland [1]. For the direct proof, see Kalman and Bertram [1]. Using the fact that in the Lyapunov equation (2.12) one may use certain positive semidefinite matrices Q as stated in Theorem 2.11, it is possible to construct different criteria structures. Converting the companion matrix to Schwarz form (Schwarz [1]) using similarity transformation (Chen and Chu [1]), it is possible to verify Hurwitz form (Kalman and Bertram [1], Parks [1], Barnett and Storey [1]), and thus the Routh criterion. The proof of the discrete table form in Theorem 2.2 is due to Chen and Chan [1]. They have produced a similarity transformation to a discrete Schwarz form as well as a proof based on (2.13). The use of Lyapunov criterion to prove Hermite and Schur-Cohn forms is due to Parks [2] and [3]. For some more details, see Parks [4]. The integral square of signals is discussed in details in Jury [1].

Chapter 3 : INTRODUCTION TO ROOT CLUSTERING

The root clustering inclusion

$$\sigma(A) \subset \aleph \qquad (3.1)$$

defined in (1.9) consists of two components — a matrix A and a region $\aleph$. As already mentioned in Chapter 1, if we insist on a simple test for (3.1), namely, algebraic inequalities, we have to restrict $\aleph$. In this chapter we describe two such restrictions. But first we define an ***algebraic region*** .

3.1 ALGEBRAIC REGIONS

We open the discussion by reviewing some concepts from set theory. A point λ is ***interior*** to a set $\aleph$, if there is an ε-neighborhood of λ containing only points of $\aleph$. λ is ***exterior*** to $\aleph$, if there is an ε-neighborhood of λ containing no points of $\aleph$. A point λ that is not interior or exterior to $\aleph$ (that is, every neighborhood of λ contains at least one point in $\aleph$ and one point not in $\aleph$), is called a boundary point. The collection of all boundary points of $\aleph$ is the ***boundary*** $\partial\aleph$. A set $\aleph$ is ***open*** if every point of $\aleph$ is an interior point. It is ***closed*** if its ***complement*** (the set $\aleph^c$ of all points not in $\aleph$) is open. The set $\aleph$ taken together with its boundary is the ***closure*** $cl(\aleph)$ of $\aleph$: $cl(\aleph) = \aleph \cup \partial\aleph$.

Let $f(x,y) \in R[x,y]$ be a real polynomial in two real variables x and y, and consider the following open region (set) in the complex plane :

$$\aleph = \{(x+iy): f(x,y) < 0\} . \qquad (3.2)$$

It is almost automatic to define the boundary of $\aleph$ as

$$\partial\aleph = \{(x+iy): f(x,y) = 0\} . \qquad (3.3)$$

However, as the following example shows, the situation is not that simple.

Example 3.1

Consider the region

$$\aleph = \{(x+iy): y^2 - (x-1)x^2 < 0\} .$$

As shown in Figure 3.1, the point (0,0) is not a boundary point, while $f(\cdot)$ vanishes both along $\partial\aleph$ and at (0,0).

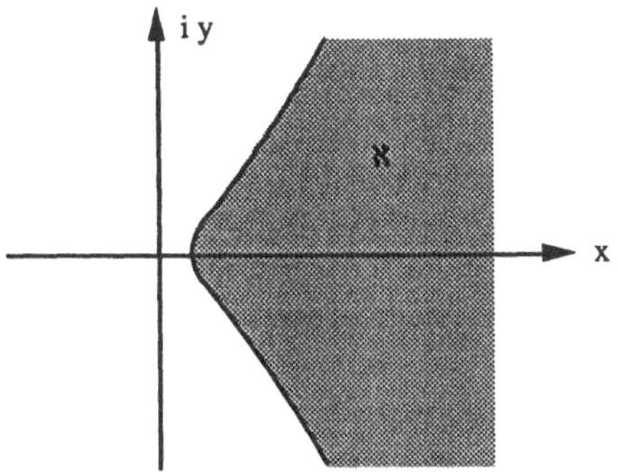

Figure 3.1 : ℵ in example 3.1

This example demonstrates that the equality $f(x,y) = 0$ may include points outside the boundary $\partial\aleph$. To exclude such singular cases, we adopt the following definition.

Definition 3.1 The region ℵ given by (3.2) is ***simple***, if the boundary $\partial\aleph$ is given by (3.3). The following is a partial characterization of a simple region.

Theorem 3.1 Consider region ℵ defined by (3.2).

(i) $\partial\aleph$ is empty, if and only if either ℵ is the empty set or ℵ is the entire complex plane C .

(ii) If $x + iy \in \partial\aleph$, then $f(x,y) = 0$.

(iii) If $f(x,y) = 0$, and $f'(x,y) \neq 0$, then $x + iy \in \partial\aleph$.

Returning to example 3.1 we see that $f(0,0) = 0$, $f(1,0) = 0$. However, $f'(0,0) = 0$ while $f'(1,0) = [-1 \quad 0] \neq 0$.

Part (iii) of the theorem is equivalent to the following corollary.

Corollary 3.1 Region ℵ given by (3.2) is simple, if $f'(x,y) \neq 0$ on $\{(x+iy): f(x,y) = 0\}$.

Proof

(i) The only regions in C, both closed and open, are C and $\varnothing$. However, by definition,

$\partial\aleph = cl(\aleph) \setminus \aleph$.

(ii) Suppose $\partial\aleph$ is not empty and take $z = x+iy$ in $\partial\aleph$. Since $z \notin \aleph$, we have $f(x,y) \geq 0$. Now, if $f(x,y) > 0$, then there exists $r > 0$ such that $f > 0$ in the ball $|z| < r$. But this is impossible, since $z \in \partial\aleph$ implies that the ball centered at z intersects ℵ. Thus, $f > 0$ is impossible, and $f \geq 0$ implies $f = 0$.

(iii) Suppose, to the contraty, that $f(z) = 0$, $f'(z) \neq 0$, but $z \in \partial\aleph$. Then, there exists $r > 0$ such that the open ball $B(z,r)$ has an empty intersection with ℵ. Since $f(\cdot)$ is differentiable,

$$f(z+h) = f(z) + f'(z)h + \|h\| \varepsilon(h)$$

with $\varepsilon(h) \to 0$ as $h \to 0$. By hypothesis, $f(z) = 0$, and $B(z,r) \cap \aleph = \emptyset$.

Thus, for all $h \in B(o,r)$, $z + h \in B(z,r)$,

$$0 \leq f(z+h) = f'(z)h + \|h\| \varepsilon(h) .$$

Using $h = t\frac{y}{\|y\|}$, we have for all $y \neq 0$, and $0 < t < r$,

$$0 \leq f'(z)\, t\frac{y}{\|y\|} + t\varepsilon\,(t\frac{y}{\|y\|}),$$

or

$$0 \leq f'(z)y + \|h\| \varepsilon(h) .$$

Taking $t \to 0$, we obtain

$$0 \leq f'(z)y .$$

Replacing y by -y, we also have :

$$0 \geq f'(z)y .$$

The last two inequalities imply

$$f'(z)y = 0 .$$

Since y is arbitrary, we finally have

$$f'(z) = 0 .$$

However $f'(z) = 0$ contradicts our hypothesis $f'(z) \neq 0$. Thus $z \in \partial\aleph$. This competes the proof.

3.2 THE STRUCTURE OF ROOT CLUSTERING TESTS

Once again, in the general root clustering problem, we have to check the truth of the sentence

$$\sigma(\Delta) \subset \aleph \qquad (3.4)$$

where $\Delta(s) = \sum_{k=0}^{n} a_k s^k$ is a given polynomial, and $\aleph$ is defined in (3.2).

If s = x+iy, we may write

$$\Delta(s) = \Delta(x+iy) = \sum_{k=0}^{n} a_k (x+iy)^k . \tag{3.5}$$

By expanding (3.5) we have:

$$\Delta(s) = \Delta_r(x, y) + i\Delta_i(x, y) \tag{3.6}$$

where $\Delta_r(\cdot)$ and $\Delta_i(\cdot)$ are polynomials in x and y. Now, $\Delta(s)$ has a zero, if and only if Δ_r and Δ_i have a common zero for some real x and y. Thus, we can replace (3.4) by checking the truth of the following ***algebraic sentence***.

> There exist no real x and y, such that
> $$\Delta_r^2(x, y) + \Delta_i^2(x, y) = 0, \qquad \text{and } f(x, y) > 0 . \tag{3.7}$$

It is known that (3.7) can be checked in a finite number of steps using ***Decision Algebra***. The basic idea is to transform the original sentence (3.7), with ***two*** variables x and y, into an equivalent one with a ***single*** variable. This last sentence can be tested using a modification on Sturm's theorem. Thus, in principle, we can solve, in a finite number of steps, the root clustering problem with respect to an ***arbitrary*** algebraic region. However, this is an unwieldy approach. First, from the structural point of view the root clustering test using Decision Algebra, has no definite structure, while the classical test, like Routh-Hurwitz, consists of ***a single set of n polynomial inequalities*** in the coefficients a_i, i = 1,2,...,n. Second, from the computational point of view, the number of steps in Decision Algebra, although finite, is very large, in comparison with the classical tests. It well may be that a numerical calculation of the roots of $\Delta(s)$ is more efficient than the use of Decision Algebra. Between the two extremes, the classical tests and Decision Algebra, there are other possibilities. The most familiar is a finite set of inequalities. In short, we distinguish between three main structures when testing root clustering:

1. A finite number of algebraic steps (Decision Algebra);
2. A finite set of polynomial inequalities;
3. A single set of polynomial inequalities.

In this book we will focus exclusively on the third structure, although in some instances we will also present the second. In order to sharpen the above distinction, we use the following definition.

Definition 3.2 A root clustering test which consists of a single set of polynomial inequalities in the coefficients $\{a_i\}$ is called ***a root clustering criterion***.

Since, in passing from Decision Algebra to a criterion, generality is lost, we expect that for some regions, a root clustering criterion cannot be found. In order to clarify this point, we present a few examples.

Example 3.2

Consider the characteristic polynomial:

$$\Delta(s) = s^2 + 2bs + a \tag{3.8}$$

and a region $\aleph$ given by

$$\aleph = \{(x + iy) : x - y^2 < 0\} \tag{3.9}$$

as shown in Figure 3.2.

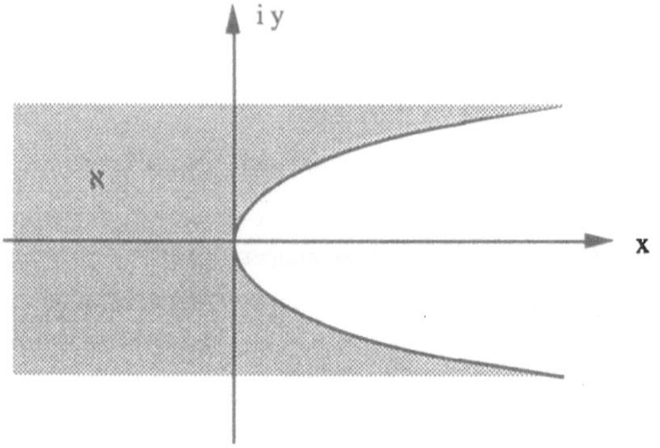

Figure 3.2 Exterior to the parabola.

Analyzing the roots of our second order $\Delta(s)$, we find that $\sigma(\Delta) \subset \aleph$, if and only if, in the parameter space R^2 spanned by a and b, $(a,b) \in \Omega$, where the region Ω is depicted in Figure 3.3.

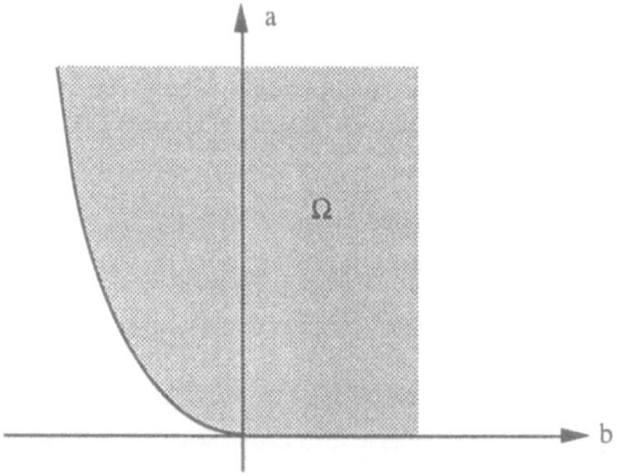

Figure 3.3 Parameter space region Ω.

The reader can easily verify that Ω in Figure 3.3 cannot be defined using a single set of polynomial inequalities in the coefficients a and b.

Example 3.3

Consider all non real numbers in the complex plane:

$$\aleph = \{(x+iy): -y^2 < 0\}\ , \tag{3.10}$$

that is, the complex plane excluding the real line. In this example, instead of a second order characteristic polynomial, we take:

$$\Delta(s) = s^4 + 2bs^2 + a\ . \tag{3.11}$$

The root clustering region in the parameter space is given in Figure 3.4. Once more, we see that Ω cannot be defined as a single set of polynomial inequalities in the coefficients a and b.

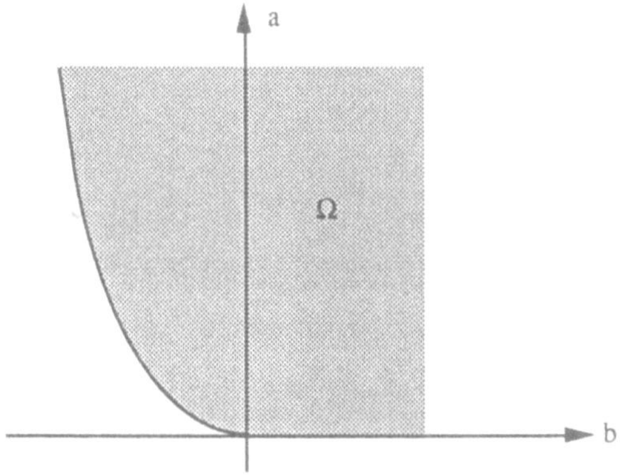

Figure 3.4 Parameter space region Ω.

The following are two more examples for which root clustering criteria do not exist.

Example 3.4

A double hyperbola with slope $a^2 > 1$:

$$\aleph = \{(x+iy): 1 - a^2x^2 + y^2 < 0\ ; \qquad a^2 > 1\}\ . \tag{3.12}$$

Example 3.5

A fourth order "parabola":

$$\aleph = \{(x+iy): x + y^4 < 0\} \tag{3.13}$$

In the following sections we start to present some root clustering criteria.

3.3 REGION'S REPRESENTATION

So far (see (1.8) and (3.2)), we have described a region in cartesian coordinates:

$$\aleph = \{(x+iy): f(x,y) < 0\} \tag{3.14}$$

where f(x,y) is the polynomial

$$f(x, y) = \sum_{i,j} f_{ij} x^i y^j . \tag{3.15}$$

Next, recall that

$$x = \frac{s + \bar{s}}{2} \qquad y = \frac{s - \bar{s}}{2i} \tag{3.16}$$

and define:

$$\phi(\alpha, \beta) = f\left(\frac{\alpha + \beta}{2}, \frac{\alpha - \beta}{2i}\right) \triangleq \sum_{i,j} \phi_{ij} \alpha^i \beta^j . \tag{3.17}$$

Then, (3.14) is equivalent to

$$\aleph = \{ s \in \mathbb{C} : \phi(s, \bar{s}) < 0 \} \tag{3.18}$$

which is the polar description of (3.14). Note that (3.17) can be written as

$$\phi(\alpha, \bar{\beta}) = L'(\alpha)\, \Phi L(\bar{\beta}) \tag{3.19}$$

where:

$$L'(\alpha) = [1 \quad \alpha \quad \alpha^2 \ldots] \tag{3.20}$$

$$\Phi = [\phi_{ij}] . \tag{3.21}$$

Fact. Φ is Hermitian: $\phi_{ij} = \bar{\phi}_{ji}$.

To prove this result, we expand $\phi(\alpha, \bar{\beta})$. For example, a second order region has the form

$$\aleph = \{ (x + iy) : f_{00} + f_{01} y + f_{02} y^2 + f_{11} xy + f_{10} x + f_{20} x^2 < 0 \} \tag{3.22}$$

Thus, a direct expansion of (3.17) implies:

$$\phi_{00} = f_{00}, \quad \phi_{11} = \tfrac{1}{2}(f_{20} + f_{02}) \tag{3.23}$$

$$\phi_{01} = \bar{\phi}_{10} = \tfrac{1}{2}(f_{10} + if_{01})$$

$$\phi_{02} = \bar{\phi}_{20} = \tfrac{1}{4}[(f_{20} - f_{02}) + if_{11}] .$$

Since Φ is hermitian it can be written $\Phi = T\Lambda T^*$, where $T^{-1} = T^*$, and Λ is diagonal with real elements δ_i. Thus,

$$\phi(\alpha, \bar{\beta}) = L'(\alpha)\Phi L(\bar{\beta}) = L'(\alpha)T\Lambda T^* L(\bar{\beta}) = \tilde{L}^*(\alpha)\Lambda\tilde{L}(\bar{\beta})$$

$$= \sum_i \delta_i \psi_i(\alpha)\overline{\psi_i(\beta)} \tag{3.24}$$

where $\psi_i(\cdot)$ are certain polynomials. Now, (3.18) takes the form:

$$\aleph = \{s \in C : \sum_i \delta_i |\psi_i(s)|^2 < 0\}. \tag{3.25}$$

This discussion reveals that it is natural to generate the polar description (3.18) from the cartesian one (3.14). Now, consider a region (3.18) where the ϕ_{ij}'s in (3.17) are ***arbitrary***. Note that (3. 18) requires that $\phi(\lambda, \bar{\lambda})$ is real. Then we have:

Theorem 3.2 Consider the two variable polynomial $\phi(\alpha, \bar{\beta}) = \sum_{i,j} \phi_{ij}\, \alpha^i \bar{\beta}^j$.

The following statements are equivalent:

(i) Φ is Hermitian.

(ii) $\phi(\alpha, \bar{\alpha})$ is real.

(iii) $\overline{\phi(\alpha, \bar{\beta})} = \phi(\beta, \bar{\alpha})$.

To prove the theorem, we need

Lemma 3.2 The complex maps $\lambda \to \lambda^i \bar{\lambda}^j$ are linearly independent.

Proof Pick complex numbers ϕ_{ij} such that $\sum_{i,j=0} \phi_{ij} \lambda^i \bar{\lambda}^j = 0$

for all $\lambda \in C$. In particular, given any $r > 0$ and any λ on the circle $|\lambda| = r$, we get, multiplying by λ^n

$$\sum_{i,j=0}^{n} \phi_{ij} r^{2j} \lambda^{n+i-j} = 0,$$

or

$$\sum_{k=0}^{2n} \left(\sum_{i=m}^{M} \phi_{i,n+i-k} r^{2(n+i-k)} \right) \lambda^k = 0,$$

where $m = \max(0,k-n)$ and $M = \min(n,k)$. Thus, for all $r > 0$ and all $k = 0,\ldots,2n$, the parenthesis vanishes, so that $\phi_{i,n+i-k} = 0$ for all $k = 0,\ldots,2n$, $i = m,\ldots,M$. Thus, $\phi_{ij} = 0$ for all $i,j = 0,1,\ldots,n$. This completes the proof of the lemma.

Proof of Theorem 3.2 We will prove the sequence (ii) ⇒ (i) ⇒ (iii) ⇒ (ii). If (ii) is satisfied, then for all $\lambda \in C$ we have

$\phi(\lambda, \bar{\lambda}) - \overline{\phi(\lambda, \bar{\lambda})} = 0$, that is, $\sum (\phi_{ij} - \bar{\phi}_{ji})\lambda^i \bar{\lambda}^j = 0$. Thus, according to Lemma 3.2, $\phi_{ij} = \bar{\phi}_{ji}$ for all i, j and (i) is satisfied. If (i) is satisfied, we have:

$$\overline{\phi(\alpha, \bar{\beta})} = \sum \bar{\phi}_{ij}\bar{\alpha}^i \beta^j = \sum \phi_{ji}\beta^j\bar{\alpha}^i = \phi(\beta, \bar{\alpha}),$$

i.e. (iii). If (iii) is satisfied, we choose $\alpha = \beta$ and obtain (ii). This completes the proof of the theorem.

Finally, if $\aleph$ is given by (3. 18) with $\Phi = \Phi^*$, use $\lambda = x + iy$ and expand $\phi(\lambda, \bar{\lambda})$ in the following way

$$\phi(\lambda, \bar{\lambda}) = f(x, y) + ig(x, y).$$

Clearly, $g(x,y) \equiv 0$ and we end with $f(x,y) < 0$, as in (3.14).

3.4 ONE VARIABLE TRANSFORMATION I

The general problem of root clustering is traced back to Hermite (1854). The family of regions introduced by him is not rich; yet, it is interesting in that it opens a possible generalization. Consider the rational function

$$w = \frac{\psi(s)}{\phi(s)} \tag{3.26}$$

where $\psi(\cdot)$ and $\phi(\cdot)$ are given polynomials. If we take (3.26) as the function $w \to s$, we can define the following region

$$\aleph = \{s \in C : |w| < 1\}$$

or, using (3.26),

$$\aleph = \{s \in C : |\psi(s)|^2 - |\phi(s)|^2 < 0\}. \tag{3.27}$$

This description of $\aleph$ is in polar coordinates. Now it is interesting to know the connection of (3.27) to the cartesian coordinates. Comparing (3.27) and (3.25), we see that they are identical, if and only if the δ_i's consist of one positive number and one negative number, and the rest are zeroes. This means that Φ has rank 2, signature 0, since rank(Φ) = p(Φ) + n(Φ), and sign(Φ) = p(Φ) - n(Φ), where p(Φ) and n(Φ) are the number of positive and negative eigenvalues of Φ, respectively. We thus conclude

Theorem 3.3 The region $\aleph$ defined by (3.14), (3.15) is identical to $\aleph$ defined by (3.27), if and only if, the matrix Φ defined by (3.17), (3.21), has rank 2, signature 0.

Now that we have established the connection between the polar and the cartesian representation of $\aleph$, we turn to the root clustering criterion. However, we first need some results from functions of matrices.

3.5 EIGENVALUES OF FUNCTIONS OF MATRICES

Let us recall a few facts from algebra. Given $A \in C^{n\times n}$, the Jordan form of A is

$$J = T^{-1}AT \tag{3.28}$$

where J is upper triangular, and the diagonal elements are the eigenvalues of A.

From (3.28), $A = TJT^{-1}$, so that

$$A^2 = TJT^{-1}TJT^{-1} = TJ^2T^{-1}$$

and in general

$$A^i = TJ^iT^{-1}. \tag{3.29}$$

J^i is upper triangular and the diagonal elements are λ_j^i, where λ_j is an eigenvalue of A.

If f(A) is a polynomial function:

$$f(A) = \sum_i f_i A^i, \tag{3.30}$$

it follows from (3.29) that

$$f(J) = T^{-1}f(A)T. \tag{3.31}$$

Next, consider the rational function

$$w(A) = \psi(A)\phi^{-1}(A) \tag{3.32}$$

where $\phi(A)$ is nonsingular. We have

$$\phi(A)w(A) = \psi(A).$$

Using (3.31)

$$T\phi(J)T^{-1}w(A) = T\psi(J)T^{-1}$$

or

$$\phi(J)T^{-1}w(A)T = \psi(J).$$

Thus

$$T^{-1}w(A)T = \psi(J)\phi^{-1}(J).$$

Since $\psi(J)$ and $\phi(J)$ are upper triangular, it follows that $\psi(J)\,\phi^{-1}(J)$ is also upper triangular and the diagonal elements are $\psi(\lambda_j)/\phi(\lambda_j)$, where:

$$\psi(\lambda) = \sum_i \psi_i\lambda^i \qquad \phi(\lambda) = \sum_i \phi_i\lambda^i. \tag{3.34}$$

Observing (3.33), we now have

Theorem 3.4 If $\lambda_1,\lambda_2,\ldots,\lambda_n$ are the eigenvalues of the matrix $A\in C^{n\text{-}n}$, and $\phi(A)$ is nonsingular, then the eigenvalues of $\psi(A)\phi^{-1}(A)$ are :

$$\frac{\psi(\lambda_1)}{\phi(\lambda_1)}, \frac{\psi(\lambda_2)}{\phi(\lambda_2)}, \ldots, \frac{\psi(\lambda_n)}{\phi(\lambda_n)}.$$

3.6 ROOT CLUSTERING CRITERIA I

We are now ready to present root clustering criteria with respect to $\aleph$ defined by (3.27).

Theorem 3.5 Consider $\aleph$ ((3.27) and Theorem 3.3) and $A \in C^{n\text{-}n}$. $\sigma(A) \subset \aleph$, if and only if $\psi(A)\phi^{-1}(A)$ has all its eigenvalues inside the unit disk.

Proof. Sufficiency — by hypothesis $\psi(A)\phi^{-1}(A)$ has all its eigenvalues inside the unit disk, (this, by the way, implies that $\phi(A)$ is nonsingular). However, according to Theorem 3.4, these eigenvalues are $\psi(\lambda_i)/\phi(\lambda_i)$, that is, $|\psi(\lambda_i)|^2/|\phi(\lambda_i)|^2 < 1$, or $|\psi(\lambda_i)|^2-|\phi(\lambda_i)|^2 < 0$.

Necessity— suppose $\sigma(A) \subset \aleph$, that is $|\psi(\lambda_i)|^2-|\phi(\lambda_i)|^2 < 0$. This implies $\phi(\lambda_i) \neq 0$. For if $\phi(\lambda_i) = 0$, then $|\psi(\lambda_i)|^2 < 0$ which is impossible. Using a new variable $w = \psi(\lambda)/\phi(\lambda)$ in (3.26), we see that $|w| < 1$. Finally, using Theorem 3.4, the eigenvalues of $\psi(A)\phi^{-1}(A)$ are all inside the unit disk. This completes the proof.

For Theorem 3.5 to be effective, it is necessary to use one of the tests in Chapter 2 for the unit disk. One way for doing so is the use of Theorem 2.9 . For $\psi(A)\phi^{-1}(A)$, equation (2.13) takes the form

$$\phi^{-1*}(A)\psi^*(A)P\psi(A)\phi^{-1}(A) - P = -Q. \tag{3.35}$$

Multiplying from left by $\phi^*(A)$ and from right by $\phi(A)$, one obtains

$$\psi^*(A)P\psi(A) - \phi^*(A)P\phi(A) = -\tilde{Q} \tag{3.36}$$

where:

$$\tilde{Q} = \phi^*(A)Q\phi(A).$$

Theorem 3.6 $\sigma(A) \subset \aleph$, defined by (3.27), if and only if given any positive definite matrix $\tilde{Q} = \tilde{Q}^*$, there exists a unique positive definite solution matrix $P = P^*$, satisfying (3.36).

Proof. A ***direct*** proof will be presented in section 6.1 for a more general case.

3.7 ONE VARIABLE TRANSFORMATION II

In the previous section we introduced a root clustering criterion to a limited class of regions. As mentioned, this is a special case of a more general theory which will be presented in a later chapter. In the present section, we introduce our first general approach to root clustering. Recall the rational function (3.26), and let:

$$w = h(s) = \frac{\psi(s)}{\phi(s)} \tag{3.37}$$

where $\psi(s)$ and $\phi(s)$ are coprime polynomials.

Now, consider the mapping $s \to w$. Clearly, each point s is mapped into a single point w. To clarify the situation, consider:

$$w = \frac{as^2 + 2bs + a}{1 - s^2}. \tag{3.38}$$

A close look reveals that the left half plane $\{(x+iy): x < 0\}$ is mapped by (3.38) into the exterior to the ellipse $\{(x+iy): \frac{x^2}{a^2} + \frac{y^2}{b^2} < 1\}$.

This leads us to define a region in the following way:

$$\aleph = C \setminus h(\,cl(C_+)\,). \tag{3.39}$$

In order to generalize (3.39) we need some mathematical preparations. Let $V(\psi(s) - w\phi(s))$ be the vanishing set of the polynomial $\psi(s) - w\phi(s)$, and let $h^{-1}(w)$ be the set of $s \in C$ such that $w = h(s)$.

Lemma 3.3 Let $h(s) = \psi(s)/\phi(s)$ be a rational complex map with $\psi(s)$ and $\phi(s)$ coprime. Then, for all $w \in C$

$$h^{-1}(w) = V(\psi(s) - w\phi(s)).$$

Proof If $s \in h^{-1}(w)$, i.e. $w = h(s)$, then $0 = \phi(s)\,(h(s)-w) = \psi(s) - w\phi(s)$ that is,

$s \in V(\psi(s) - w\phi(s))$. Conversely, if $\psi(s) - w\phi(s) = 0$, then $\phi(s) \neq 0$ (otherwise, $\psi(s)$ also vanishes and $\psi(s)$ and $\phi(s)$ have a common factor). Thus, $w = h(s)$ so that $s \in h^{-1}(w)$.

Now, let Ω be a nonempty subset of C and define the following region in the complex plane.

$$\aleph = C \setminus h(C \setminus \Omega). \tag{3.40}$$

Note that if we take $\Omega = C_-$, we get (3.39) as a special case.
To characterize $\aleph$, we state our first main result.

Theorem 3.7 Let $w = h(s) = \psi(s)/\phi(s)$ be a rational map with $\psi(s)$ and $\phi(s)$ coprime. Let Ω be a nonempty subset of C. Then the following are equivalent.

(i) $\aleph = C \setminus h(C \setminus \Omega)$;

(ii) $\aleph = \{w \in C : h^{-1}(w) \subset \Omega\}$;

(iii) $\aleph = \{w \in C : V(\psi(s) - w\phi(s)) \subset \Omega\}$.

Proof (ii) $\Leftrightarrow$ (iii) is clear from Lemma 3.3. We prove (i) $\Leftrightarrow$ (ii).
(ii) $\Rightarrow$ (i) : If $w \in h(C \setminus \Omega)$, i.e. $w = h(s)$ for some $s \notin \Omega$, then $h^{-1}(w) \not\subset \Omega$. Thus, by contraposition we have $w \in C \setminus h(C \setminus \Omega)$ if $h^{-1}(w) \subset \Omega$.
(i) $\Rightarrow$ (ii): Assume that $w \notin C \setminus h(C \setminus \Omega)$ and take $s \in h^{-1}(w)$, i.e. $w = h(s)$. Since $w \notin h(C \setminus \Omega)$ we have $s \notin \Omega$. Thus $h^{-1}(w) \subset \Omega$.

Remarks

1. $\aleph$ in (3.40) is far more general than (3.39). First we do not restrict $h(\cdot)$. Second, Ω is not restricted to C_-.
2. To characterize $\aleph$, let $w = x+iy$ and define

$$g(s) = \psi(s) - (x+iy)\phi(s) . \tag{3.41}$$

Suppose that Ω has a known criterion for $V[g(s)] \subset \Omega$. In this case we generate the respective set of inequalities

$$\Delta_j(x, y) > 0 \qquad j = 1, 2, \dots .$$

Then

$$\aleph \bigcap_j \{(x+iy): \Delta_j(x, y) > 0\} . \tag{3.42}$$

3. Note that the boundary $\partial\aleph$ in (3.39) is a ***rational curve*** . On the other hand, if we choose Ω in (3.40) properly, we may have ***irrational boundary*** . As an illustration, take:

$$\Omega = \{(x+iy): -1 + x^4 + y^4 < 0\} . \tag{3.43}$$

The boundary of (3.43) is irrational. But according to Example 4.5, this region has a root clustering criterion. Thus, using (3.40), we can generate a region with irrational boundary.

4. Our admissible region is defined by (3.40), generated by an arbitrary (minimal) rational function $h(\cdot)$, operating on a region having a known root clustering criterion.

Example 3.6 Consider

$$h(s) = (a+ib)s + c .$$

Using (3.41)

$$g(s) = (a+ib)s + c-(x+iy) .$$

Take $\Omega = C_-$. Then $g(s)$ is Hurwitz implies the ***half*** plane

$$\aleph = \{(x+iy): ax+by-ac < 0\} . \tag{3.44}$$

Example 3.7 Consider

$$h(s) = as^2 + s + b .$$

Using (3.41)

$$g(s) = as^2 + s + b - (x+iy) .$$

Take $\Omega = C_{-}$. Then g(s) is Hurwitz implies the ***parabola***

$$\aleph = \{(x+iy): ay^2+x-b < 0\} .$$

Example 3.8 Consider

$$h(s) = \frac{as^2 + 2bs + c}{s^2 + 1} .$$

Using (3.41)

$$g(s) = as^2 + 2bs + c - (x + iy)(s^2 + 1)$$

$$= (a - x - iy)s^2 + 2bs + (c - x - iy) .$$

Take $\Omega = C_{-}$. Then, g(s) is Hurwitz, if and only if

$$\Delta_1(x, y) = b(a - x) > 0$$

$$\Delta_2(x, y) = 4b^2(a - x)(c - x) - (c - a)^2 y^2 > 0 .$$

Choose $a = -c > 0, b > 0$ and (3.42) becomes the ***left*** hyperbola

$$\aleph = \{(x + iy): 1 - \frac{x^2}{a^2} + \frac{y^2}{b^2} < 0) \cap \{(x + iy): x < 0) . \tag{3.46}$$

Example 3.9 Consider

$$h(s) = \frac{as^2 + 2bs + c}{s^2 - 1} .$$

Using (3.41)

$$g(s) = as^2 + 2bs + c - (x + iy)(s^2 - 1)$$

$$= (a - x - iy)s^2 + 2bs + (c + x + iy) .$$

Take $\Omega = C_{-}$, then g(s) is Hurwitz, if and only if

$$\Delta_1(x, y) = b(a - x) > 0$$

$$\Delta_2(x, y) = 4b^2(a - x)(c + x) - (c + a)^2 y^2 > 0 .$$

Choose $a = c > 0$, $b > 0$, and (3.42) becomes the ***ellipse***

$$\aleph = \{(x+iy): -1 + \frac{x^2}{a^2} + \frac{y^2}{b^2} < 0\}. \tag{3.47}$$

Example 3.10 Consider

$$h(s) = \frac{as^2}{s+1}.$$

Using (3.41)

$$g(s) = as^2 - (x+iy)(s+1)$$

$$= as^2 - (x+iy)s - (x+iy).$$

Take $\Omega = C_-$. Then $g(s)$ is Hurwitz if and only if

$$\Delta_1(x,y) = -\frac{x}{y} > 0$$

$$\Delta_2(x,y) = -\frac{y^2}{a^2} - \frac{x}{a}\left(\frac{y^2}{a^2} + \frac{x^2}{a^2}\right) > 0.$$

This implies

$$\aleph = \{(x+iy): x^3 + (x+a)y^2 < 0\}. \tag{3.48}$$

We can conclude that each h(s) and Ω generate $\aleph$. However, we do not have a way to generate h(s) for a given $\aleph$.

3.8 ROOT CLUSTERING CRITERIA II

Given a polynomial $\Delta(z)$, we call its ***vanishing set*** (the set of all roots) the spectrum $\sigma(\Delta)$. We now state a root clustering theorem.

Theorem 3.8 Consider region $\aleph$ defined by $\aleph = C\backslash h(C\backslash\Omega)$ with $h(s) = \psi(s)/\phi(s)$ minimal and Ω a nonempty region in the complex plane. Let $\Delta(z) = \sum_{i=0}^{n} a_i z^i$ be a complex polynomial. Then, $\sigma(\Delta) \subset \aleph$ if and only if $\tilde{\Delta} \triangleq \Delta(\psi(s)/\phi(s)) = \sum_{i=0}^{n} a_i \psi^i(s)\phi^{n-i}(s)$ satisfies $\sigma(\tilde{\Delta}) \subset \Omega$.

Proof Necessity — we assume that $\Delta(z)$ does not vanish in $C\backslash\aleph$, and have to prove that $\sigma(\tilde{\Delta}) \subset \Omega$. Suppose on the contrary, that $\tilde{\Delta}(s)$ vanishes at some points in

$C\backslash\Omega$. Then $\phi(s) \neq 0$ at these points, for if $\phi(s) = 0$, we must have $\psi(s) = 0$ and we violate minimality. Now, divide $\tilde{\Delta}(s)$ by $\phi^n(s)$ and we have $\Delta(\psi(s)/\phi(s)) = 0$. However, any $s \notin C\backslash\Omega$ has an image in $C\backslash\aleph$ by $h(\cdot)$. Thus $\Delta(z) = 0$ in $C\backslash\aleph$. This contradicts the hypothesis. Sufficiency — suppose $\sigma(\tilde{\Delta}) \subset \Omega$. Then $\tilde{\Delta}(s) \neq 0$ in $C\backslash\Omega$. Suppose, contrary to the theorem, that $\Delta(z_0) = 0$ for some $z_0 \in C\backslash\aleph$. By construction, there exists $s_0 \in C\backslash\Omega$ such that $z_0 = \psi(s_0)/\phi(s_0)$. Thus, $\Delta(\psi(s_0)/\phi(s_0)) = 0$, and for all finite z_0, $\phi(s_0) \neq 0$. Thus, $\tilde{\Delta}(s_0) = 0$. This contradicts the hypothesis.

Corollary 3.8 For $\Omega = C_-$, we have $\aleph = C\backslash h(\bar{C}_+)$. Then $\sigma(\Delta) \subset \aleph$ if and only if $\tilde{\Delta}(s)$ is Hurwitz.

Example 3.11 Let $\aleph$ be the left hyperbola

$$\aleph = \{(x+iy): -1 + x^2 - y^2 < 0\} \cap \{(x+iy): x < 0\}$$

and consider

$$\Delta(z) = z^2 + 5z + 6.$$

To test $\sigma(\Delta) \subset \aleph$, recall Example 3.8. Then, take $a = -c = b = 1$, so that

$$h(s) = \frac{s^2 + 2s - 1}{s^2 + 1}.$$

Thus,

$$\tilde{\Delta}(s) = (s^2 + 2s - 1) + 5(s^2 + 2s - 1)(s^2 + 1) + 6(s^2 + 1)$$

$$= 3s^4 + 7s^3 + 4s^2 + 3s + 1.$$

It is easy to verify that $\tilde{\Delta}(s)$ is Hurwitz, so that all the roots of $\Delta(z)$ are clustered in the above hyperbole.

Next, we present an optimality property of the root clustering criterion presented in Theorem 3.8.

Theorem 3.9 Let $h(\cdot)$, Ω, $\Delta(\cdot)$, and $\tilde{\Delta}(\cdot)$ be defined as in Theorem 3.8. Then, $\sigma(\Delta) \subset W \Leftrightarrow \sigma(\tilde{\Delta}) \subset \Omega$ for all nonconstant complex polynomials $\Delta(\cdot)$, implies that $W = C\backslash h(C\backslash\Omega) = \aleph$.

Proof Take $\Delta(z) = z - w$ with $w \in C$. If $w \in W$ then $\sigma(\Delta) \subset W$ from where $\sigma(\tilde{\Delta}) \subset \Omega$, i.e., since we choose $\Delta(z) = z - w$, we have $\tilde{\Delta}(s) = \psi(s) - w\phi(s)$, so that $V(\psi - w\phi) \subset \Omega$ and $w \in \aleph$. Conversely, if $w \in \aleph$ then $\sigma(\tilde{\Delta}) \subset \Omega$ which implies $\sigma(\Delta) \subset W$, i.e. $w \in W$ since $\Delta(z) = z - w$.

We now present an important generalization to our previous results. Previously, in (3.37), we focused on:

$$\phi(s)w - \psi(s) = 0 .$$

Now, consider the two-variable polynomial

$$g(w, s) = \sum_{i=0}^{m} g_i(s)w^i \tag{3.49}$$

where $\{g_i(s)\}$ is coprime. This means that $g_i(s)$, $i = 1,2,...,m$, do not have a common factor.Motivated by (3.40) we define

$$\aleph = C \backslash Vg(w, C \backslash \Omega) . \tag{3.50}$$

Then, Theorem 3.7 is generalized to the following theorem.

Theorem 3.10 Let $g(w, s) = \sum_{i=0}^{m} g_i(s)w^i$ be a two variable polynomial such that $\{g_i(s)\}$ is coprime. Let Ω be a nonempty set of C. Then the following are equivalent.

(i) $\aleph = C \backslash Vg(w, C \backslash \Omega)$,

(ii) $\aleph = \{w \in C : Vg(w,s) \subset \Omega\}$.

Thus, if we set $w = x+iy$ and use a criterion with respect to Ω, we obtain a characterization for $\aleph$. Finally, the root clustering criterion has the following form.

Theorem 3.11 Let $g(w, s) = \sum_{i=0}^{m} g_i(s) w^i$ be a two variable polynomial with $\{g_i(s)\}$ coprime, and consider $\aleph$ given by (3.50) and Theorem (3.10).

Let $\Delta(z) = \sum_{i=0}^{n} a_i z^i$ be a complex polynomial and define $\tilde{\Delta}(s) = \text{Res}\,[\Delta(z), \sum_{i=0}^{m} g_i(s) z^i]$, where "Res" is the ***resultant*** with respect to z. Then, $\sigma(\Delta) \subset \aleph$ if and only if $\sigma(\tilde{\Delta}) \subset \Omega$.

The proofs of these theorems can be obtained in the same ways as in Theorems 3.7 and 3.8. Later, in section 5.1, a polynomial of the form $\tilde{\Delta}(s)$ is called a ***composite polynomial***.

There are three popular ways to calculate the resultant of two polynomials. The first is based on the companion matrix, the second on the Sylvester matrix, and the third on the Bezoutian matrix, the details of which are presented in Section 5.1. It is important to recall, at this point, that if two regions have root clustering criteria, so has their ***intersection*** , which consists of the ***union*** of the ***respective criteria***. Observing g(w,s) above, we see that if this two-variable polynomial can be written as a product of linear factors $g(w, s) = \prod_i (\phi_i(s)w - \psi_i(s))$, then we do not need our generalization at all. In that case, Theorem 5 reduces to the intersection of regions, as mentioned above. Thus, ***we are interested in cases in which $g(w,s)$ does not have linear factors in w***.

NOTES AND REFERENCES

Simple regions are taken from Chojnowski [1] and Gutman and Chojnowski [1]. Decision Algebra was developed by Tarski [1]. In an interesting article, Anderson, Bose and Jury [1] point out a way to use Decision Algebra for solving the root clustering problem for an arbitrary algebraic region. Counterexamples 3.2 - 3.5 are taken from Chojnowski [1]. The region's representation in Section 3.3 is originated in Gutman [1]. Theorem 3.2 is due to Gutman and Chojnowski [1]. The class of regions (3.27) is implied by Hermite [1]. Eigenvalues of functions of matrices can be found in many texts, like Lancaster and Tismenetsky [1]. Theorems 3.5 and 3.6 can be found in Barnett [2] and Gutman and Jury [1], but see also Kalman [1]. Section 3.7 is due to Chojnowski and Gutman [1]. First steps in this direction were taken by Walach and Zeheb [1], Zeheb and Hertz [1], Barnett and Scraton [1], and Sondergeld [1].

Chapter 4: TRANSFORMABLE REGIONS

In this chapter we start a systematic investigation of root clustering. Step by step, we construct three families of regions for which root clustering criteria are possible. These regions are called ***transformable*** . We will show the connection of these regions to root clustering; however, the complete criteria will be presented only in Chapter 5.

4.1 P-TRANSFORMABILITY

Because of their importance, we recall some definitions from Section 3.3. Consider the region

$$\aleph = \{(x+iy): f(x,y) < 0\} \tag{4.1}$$

where f(x,y) is the polynomial

$$f(x, y) = \sum_{i,j} f_{ij} x^i y^j . \tag{4.2}$$

If $s = x+iy$, we may write

$$x = \frac{s+\bar{s}}{2} , \qquad y = \frac{s-\bar{s}}{2i} . \tag{4.3}$$

Define

$$\phi(s, \bar{s}) = f\left(\frac{s+\bar{s}}{2} , \frac{s-\bar{s}}{2i}\right). \tag{4.4}$$

Then, the region $\aleph$ in (4.1) is defined by

$$\aleph = \{s \in \mathbb{C} : \phi(s,\bar{s}) < 0\}. \tag{4.5}$$

Motivated by the structure of $\mu(s,s)$, we define the following ***two-variable polynomial***

$$\begin{aligned}\phi(\alpha, \bar{\beta}) = f\left(\frac{\alpha+\bar{\beta}}{2} , \frac{\alpha-\bar{\beta}}{2i}\right) &\triangleq \sum_{i,j} \phi_{ij} \alpha^i \bar{\beta}^j , \\ &\triangleq L'(\alpha)\, \Phi L(\bar{\beta}) ,\end{aligned} \tag{4.6}$$

where

$$L'(\alpha) = [\, 1 \quad \alpha \quad \alpha^2 \; \dots \,] , \tag{4.7}$$

$$\Phi = [\phi_{ij}] . \tag{4.8}$$

Based on Theorem 3.2, we have

Theorem 4.1 In (4.6) ,

$$\text{(i)} \quad \phi_{ij} = \bar{\phi}_{ji} \ ,$$

$$\text{(ii)} \quad \overline{\phi(\alpha, \bar{\beta})} = \phi(\beta, \bar{\alpha}) \ . \tag{4.9}$$

Next, we investigate the function $\mu(\alpha, \bar{\beta})$ for two classical regions. For the ***left half plane***

$$f(x,y) = x$$

so that

$$\phi(\alpha, \bar{\beta}) = \frac{\alpha + \bar{\beta}}{2} \ .$$

It is clear that if $\mathrm{Re}[\alpha] < 0$ and $\mathrm{Re}[\beta] < 0$, then

$$\mathrm{Re}[\phi] = \frac{1}{2}\,\mathrm{Re}[\alpha, \bar{\beta}] < 0 \ .$$

In other words, the function $\frac{\alpha + \bar{\beta}}{2}$ maps the left half plane into itself.

For the ***unit disk***

$$f(x,y) = -1+x^2+y^2$$

so that

$$\phi(\alpha, \bar{\beta}) = -1 + \alpha\bar{\beta} \ .$$

If $|\alpha| < 1$ and $|\beta| < 1$, then $|1+\mu| < 1$, and $\mathrm{Re}[\phi] < 0$. In other words, the function $-1 + \alpha\bar{\beta}$ maps the unit disk into the shifted unit disk contained in the left half plane.

The above two simple examples motivate us to define the concept of ***transformability***.

Definition 4.1 A region $\aleph$ defined in (4.1) is ***P-transformable*** (with respect to the left half plane), if for $\phi(\cdot)$ defined in (4.6),

$$\text{all } \alpha, \beta \in \aleph \Rightarrow \mathrm{Re}[\phi(\alpha, \bar{\beta})] < 0 \ . \tag{4.10}$$

Definition 4.2 A region $\mathrm{cl}(\aleph)$, where $\aleph$ is given in (4.1) is ***P-transformable*** (with respect to the left half plane), if for $\phi(\cdot)$ defined in (4.6),

$$\text{all } \alpha, \beta \in \mathrm{cl}(\aleph) \Rightarrow \mathrm{Re}[\phi(\alpha, \bar{\beta})] \le 0 \ . \tag{4.11}$$

Theorem 4.2 Definitions 4.1 and 4.2 are equivalent.

In other words, the open region $\aleph$ is transformable, if and only if the closure $cl(\aleph)$ is transformable. As we will see, (4.11) has a computation advantage.

Proof (4. 10) $\Rightarrow$ (4. 11) by continuity of $(\alpha, \beta) \rightarrow Re[\phi(\alpha, \bar{\beta})]$. We prove (4. 11) $\Rightarrow$ (4.10) by a contradition. Suppose (4.11) is satisfied, but (4.10) is not; that is, $Re[\phi(\alpha_0, \bar{\beta}_0)] = 0$ for some $\alpha_o, \beta_o \in \aleph$. We can write $\aleph = \bigcup_{i \in I} \aleph_i$, where the $\aleph_i$'s are the connected components of $\aleph$. Since $\aleph$ is open, each $\aleph_i$ is open and thus a domian. The polynomial function $\phi(\cdot\,, \bar{\beta}_0)$ is analytic in the domain $\aleph_{io}$ containing α_0, and $\alpha \rightarrow Re[\phi(\alpha, \bar{\beta}_o)]$ has a maximum at $\alpha = \alpha_o$. But, according to the maximum principle, if Ω is a domain and $h(\cdot)$ is analytic in Ω, then $Re(h)$ does not have a local maximum in Ω, unless $h(\cdot)$ is constant. Thus

$$\phi(\alpha, \bar{\beta}_o) \sum_{i,j} \phi_{ij} \alpha^i \bar{\beta}_o^j = \sum_i \left(\sum_j \phi_{ij} \bar{\beta}_o^j \right) \alpha^i$$

is constant in $\aleph_{io}$. Since it vanishes in $\aleph_{io}$, we must have $\sum_j \phi_{ij} \bar{\beta}_o^j = 0$ for all i .

Finally, considering $\beta_0 \in \aleph$, we have

$$0 > \phi(\beta_o, \beta_o) = \sum_i (\sum_j \phi_{ij} \bar{\beta}_o^j) = 0$$

a contradiction. This completes the proof.

Before concentrating on a test for transformability, we wish to demonstrate the benefit of introducing such a concept.

Theorem 4.3 Let $\aleph$ be a P-transformable region, and let $\Delta(\lambda) = \sum_{i=o}^{n} a_i \lambda^i$, $a_n > 0$, be a complex polynomial with roots $\{\lambda_i\}$. Let $q(\eta) = \sum_{i=o}^{n^2} q_i \eta^i$, $q_n > 0$, be the real polynomial with roots $\{\phi(\lambda_i, \bar{\lambda}_j)\}$, $i, j = 1, 2, \ldots, n$, where $\phi(\lambda_i, \bar{\lambda}_j)$ is given by (4.6). For the roots of $\Delta(\lambda)$ to lie in $\aleph$ ($\sigma(\Delta) \subset \aleph$), it is necessary and sufficient that $q_i > 0$, $i = 1,2,\ldots,n^2$.

Proof We first prove that $q(\mu)$ is a real polynomial. According to Theorem 4.1, $\phi(\lambda_i, \bar{\lambda}_j) = \phi(\lambda_j, \bar{\lambda}_i)$.Since $i, j = 1, 2, \ldots, n$, for each index pair (i, j), there exists a pair (j,i). Thus the roots of $q(\mu)$ can be arranged in conjugate complex pairs. Hence $q(\eta)$ is a real polynomial—all its coefficients are real. Sufficiency—$q_i > 0$ $\forall$ i implies that $q(\eta)$ has only negative real roots. (Suppose the opposite. Then $q(\eta) > 0$ so μ cannot satisfy $q(\eta) = 0$). These real roots include $\phi(\lambda_i, \bar{\lambda}_i) < 0$, for each λ_i complex or real. Finally, (4.5) implies $\lambda_i \in \aleph$ $\forall$ i. Necessity — suppose $\lambda_i \in \aleph$ $\forall$ i. Since $\aleph$ is transformable, it follows that $Re[\eta] < 0$. If $-a_j$ and $-b_k + ic_k$ are real and complex roots of $q(\mu) = 0$, respectively, we can write

$$q(\eta) = \prod_j (\eta + a_j) \prod_k (\eta + b_k + ic_k)(\eta + b_k - ic_k)$$

$$= \prod_j (\eta + a_j) \prod_k [(\eta + b_k)^2 + c_k^2] ; \qquad a_j > 0, \quad b_k > 0.$$

The last equation results in a polynomial with positive coefficients, that is, $q_i > 0 \; \forall \, i$. This completes the proof.

Note that our interest in transformable regions lies in the possibility to state root clustering criteria. In order to apply Theorem 4.3, we need a way to generate the polynomial $q(\eta)$ given $\Delta(\lambda)$ and $\aleph$. This will be the subject of the next chapter. In Theorem 4.3 we are concerned with the inclusion $\sigma(\Delta) \subset \aleph$. If we are interested in $\sigma(\Delta) \subset cl(\aleph)$, we see that in the necessity part of the proof, $q_i \geq 0$ replaces $q_i > 0$. In the sufficiency part, we obtain $\phi(\lambda_i, \lambda_i) \leq 0$. However, it is possible that $f(x,y) = 0$ and $x+iy \notin cl(\aleph)$, as in Example 3.1. In order to exclude such a case, we have to assume that $\aleph$ is ***simple*** (Definition 3.1).

Theorem 4.4 Let $\aleph$ be a P-transformable simple region, and let $\Delta(\lambda) = \sum_{i=0}^{n} a_i\lambda^i$, $a_n > 0$, be a complex polynomial with roots $\{\lambda_i\}$. Let $q(\eta) = \sum_{i=0}^{n^2} q_i\eta^i$, $q_n > 0$, be the real polynomial with roots $\{\phi(\lambda_i, \bar{\lambda}_j)\}$, $i, j = 1, 2, \ldots, n$, where $\phi(\lambda_i, \bar{\lambda}_j)$ is given by (4.6). $\sigma(\Delta) \subset cl(\aleph)$, if and only if $q_i \geq 0$, $i = 1,2,\ldots,n^2$.

In Theorems 4.3 and 4.4 the number of the inequalities is n^2. We see that the present form of root clustering theory calls for more inequalities than the classical theory does. In a later chapter we will discuss the computational aspect of this observation. At this stage, for the case of real polynomials, we can reduce the number of inequalities from n^2 to $n(n+1)/2$. First, observe that if $\Delta(\lambda)$ is real, its roots appear in conjugate complex pairs. Thus, it is sufficient to discuss root clustering of symmetric regions (in the complex plane) with respect to the real axis. If $\aleph$ is symmetric (with respect to the real axis), then j in (4.2) is even.

Thus $y^j = \left(\frac{s - \bar{s}}{2i}\right)^j$ is free of i.

As a result, the matrix C in (4.6) becomes ***symmetric*** .

Theorem 4.5 If $\aleph$ is symmetric, then in (4.6)

(i) ϕ is symmetric: $\phi_{ij} = \phi_{ji}$,

(ii) $\phi(\alpha,\beta) = \phi(\beta,\alpha)$.

Note an important change. Here, we use $\phi(\alpha,\beta)$ rather than $\phi(\alpha,\beta)$. Thus, if we consider $\phi(\lambda_i,\lambda_j)$ for all $\{i,j\}$, we can make the following decomposition :

$$\prod_{i,j=1}^{n} (\eta - \phi(\lambda_i, \lambda_j)) = \prod_{i=1}^{n}(\eta - \phi(\lambda_i, \lambda_i))[\prod_{1 \leq j < i \leq n} (\eta - \phi(\lambda_i, \lambda_j))]^2 . \tag{4.12}$$

The reader will not find it difficult to prove that each component of this decomposition is a real polynomial.

Now, similar to Definitions 4.1, 4.2 we make the following definition.

Definition 4.3 A symmetric region $\aleph$ defined in (4.1) is ***P-transformable***, if

(i) all $\alpha,\beta \in \aleph \Rightarrow \mathrm{Re}[\phi(\alpha,\beta)] < 0$, or equivalently

(ii) all $\alpha,\beta \in \aleph \Rightarrow \mathrm{Re}[\phi(\alpha,\beta)] \leq 0$.

Similar to Theorems 4.3 and 4.4 we have

Theorem 4.6 Let $\aleph$ be a symmetric P-transformable region, and let $\Delta(\lambda) = \sum_{i=o}^{n} a_i\lambda^i$, $a_n > 0$, be a real polynomial with roots $\{\lambda_i\}$. Let $q(\eta) = \sum_{i=o}^{m} q_i\eta^i$, $q_m > 0$, $m = \frac{1}{2}n(n+1)$, be the real polynomial with roots $\{\phi(\lambda_i, \lambda_j)\}$, $1 \leq j \leq i \leq n$. Then, $\sigma(\Delta) \subset \aleph$, if and only if $q_i > 0$. In particular, if and only if in the polynomials

$$\text{(i)} \quad \prod_{i=1}^{n} (\eta - \phi(\lambda_i, \lambda_i)),$$

$$\text{(ii)} \quad \prod_{1 \leq j < i \leq n} (\eta - \phi(\lambda_i, \lambda_j)).$$

all the coefficients of η are positive.

Theorem 4.7 As Theorem 4.6, but

(i) $\aleph$ is a symmetric, simple and P-transformable,

(ii) $\sigma(\Delta) \subset \mathrm{cl}(\aleph)$, if and only if $q_i \geq 0$.

Yet, in some cases, we can make a further simplification. Take, for example, the LHP. Here, $\phi(\lambda_i, \lambda_j) = \frac{1}{2}(\lambda_i + \lambda_j)$. Thus, $\phi(\lambda_i + \lambda_i) = \lambda_i$, and in Theorem 4. 6, polynomial (i) takes the form $\prod_{i=1}^{n} (\eta - \lambda_i) = \Delta(\eta)$, which is the given polynomial. We now replace condition (i) in the theorem, namely, $a_i > 0 \;\; \forall\, i$, by $a_0 > 0$, and show that the sufficiency holds. If $a_o = \prod_i (-\lambda_i) > 0$, then $\Delta(\lambda)$ has either an even number of positive real roots or no such roots, but ***never*** a single positive real root. Now suppose $\Delta(\lambda)$ has (an even number of) positive real roots. Then each pair generates a positive $\phi(\lambda_i,\lambda_j)$. But this is impossible, by condition (ii) of the theorem. Thus, all of the real roots of $\Delta(\lambda)$ are negative. Condition (ii) of the theorem guarantees that all the complex roots of $\Delta(\lambda)$ have real parts. In conclusion, we have

Theorem 4.8 As Theorem 4.6 but $\aleph$ is the LHP, and condition (i) is replaced by

(i)' $a_0 > 0$.

In a similar way it is possible to prove a unit disk version.

Theorem 4.9 As Theorem 4.6, but $\aleph$ is the UD, and condition (ii) is replaced by

(ii) $\Delta(1) > 0$, and $(-1)^n \Delta(-1) > 0$.

As a result, the number of root clustering inequalities reduces for the left half plane and the unit disk, to $\frac{1}{2} n(n-1)+1$ and $\frac{1}{2} n(n-1)+2$, respectively.

After establishing a motivation for the notion of transformability, we are ready to discuss a test for transformability.

4.2 A TEST FOR P-TRANSFORMABILITY

Recall that a region is transformable, if $\alpha, \beta \in \aleph \Rightarrow \mathrm{Re}[\phi(\alpha, \bar{\beta})] < 0$. We expand $\mathrm{Re}[\phi(\alpha, \bar{\beta})]$ in the following way:

$$2\,\mathrm{Re}[\phi(\alpha, \bar{\beta})] = \phi(\alpha, \bar{\beta}) + \phi(\beta, \bar{\alpha}).$$

Thus,

$$\begin{aligned} 2\,\mathrm{Re}[\phi(\alpha, \bar{\beta})] - \phi(\alpha, \bar{\alpha}) - \phi(\beta, \bar{\beta}) &= (\phi(\alpha, \bar{\beta}) - \phi(\alpha, \bar{\alpha})) + (\phi(\beta, \bar{\alpha}) - \phi(\beta, \bar{\beta})) \\ &= \sum_{ij} \phi_{ij}\alpha^i(\bar{\beta}^j - \bar{\alpha}^j) + \sum_{ij} \phi_{ij}\beta^i(\bar{\alpha}^j - \bar{\beta}^j) \\ &= -\sum_{ij} \phi_{ij}(\alpha^i - \beta^i)(\bar{\alpha}^j - \bar{\beta}^j). \end{aligned} \tag{4.13}$$

Next, for $p \ge 0$, we define the following symmetrical polynomial

$$S_p(\alpha, \beta) = \frac{\alpha^p - \beta^p}{\alpha - \beta}. \tag{4.14}$$

In particular

$$S_0(\alpha,\beta) = 0$$
$$S_1(\alpha,\beta) = 1$$
$$S_2(\alpha,\beta) = \alpha+\beta$$
$$S_3(\alpha,\beta) = \alpha^2+\alpha\beta+\beta^2$$
$$S_4(\alpha,\beta) = \alpha^3+\alpha^2\beta+\alpha\beta^2+\beta^3.$$

We also define

$$\delta(\alpha, \beta) = \sum_{i,j} \phi_{ij} S_i(\alpha, \beta)\, S_j(\bar{\alpha}, \bar{\beta}) . \tag{4.15}$$

Substituting (4.15) in (4.13), yields:

$$2\,\mathrm{Re}[\phi(\alpha, \bar{\beta})] = \phi(\alpha, \bar{\alpha}) + \phi(\beta, \bar{\beta}) - |\alpha - \beta|^2\, \delta\alpha, \beta) . \tag{4.16}$$

The importance of (4. 16) can be seen in the separation of $\phi(\alpha, \bar{\alpha})$ and $\phi(\beta, \bar{\beta})$ from the rest. Recall that $\alpha, \beta \in \aleph \Rightarrow \phi(\alpha, \bar{\alpha}) < 0,\ \ \phi(\beta, \bar{\beta}) < 0$. Moreover, $\delta(\alpha, \beta)$ is a polynomial whose degree is less than that of $\mathrm{Re}[\ \phi(\alpha, \bar{\beta}]$. In the following, we present several alternatives to test transformability.

Theorem 4.10 Let a region $\aleph$ in the complex plane be given by (4.1), (4.2).

(i) A ***sufficient*** condition for $\aleph$ to be P-transformable is $\delta(\alpha,\beta) \geq 0\ \ \forall\, \alpha,\beta \in \aleph$,

(ii) A ***necessary*** condition for $\aleph$ to be P-transformable is $\delta(\alpha,\beta) \geq 0$, and $\forall\, \alpha,\beta \in \partial\aleph$, $\alpha \neq \beta$, and $\delta(\alpha,\beta) \geq 0$, $\forall\, \alpha \in \partial\aleph\backslash I$, where I is the set of isolated points of the boundary $\partial\aleph$.

Proof (i) for $\alpha, \beta \in \mathrm{cl}(\aleph)$ we have $\phi(\alpha, \bar{\alpha}) \leq 0$, and $\phi(\beta, \bar{\beta}) \leq 0$. Thus $2\,\mathrm{Re}[\ \phi(\alpha, \bar{\beta})] \leq -\,|\alpha - \beta|^2\, \delta(\alpha, \beta)$. If $\delta(\alpha, \beta) \geq 0\ \ \ \forall\, \alpha, \beta \in \partial\aleph$, $\alpha \neq \beta$, then clearly $2\,\mathrm{Re}[\ \phi(\alpha, \bar{\beta})] \leq 0$. (ii) Recall that $\alpha, \beta \in \partial\aleph \Rightarrow \phi(\alpha, \bar{\alpha}) = \phi(\beta, \bar{\beta}) = 0$. Thus, for $\aleph$ to be transformable, it is necessary that $2\,\mathrm{Re}[\phi(\alpha, \bar{\beta})] = -\,|\alpha - \beta|^2\, \delta(\alpha, \beta)\ \forall\, \alpha, \beta \in \partial\aleph$. This implies $\delta(\alpha, \beta) \geq 0$ if $\alpha \neq \beta$, and if α is not an isolated point of $\partial\aleph$, $\delta(\alpha,\alpha_n) \geq 0$ where $\{\alpha_n\}$ is a sequence in $\partial\aleph$ convergingto α as $\eta \to \infty$. Finally, by continuity $\delta(\alpha,\alpha) \geq 0$. This completes the proof.

With the aid of Theorem 4.10, we can be more specific.

Corollary 4.10

(i) Any first order region is P-transformable.

(ii) A second order region is P-transformable, if and only if $f_{02} + f_{20} \geq 0$.

(iii) A third order region is P-transformable, if $f_{02} + f_{20} \geq 0$ and $f_{12} + 3f_{30} = f_{21} + 3f_{03} = 0$.

Proof

(i) In this case, $\delta(\alpha,\beta) = 0\ \ \ \forall\, \alpha,\beta$.

(ii) Here, $\delta(\alpha, \beta) = \frac{1}{2}(f_{02} + f_{20})\ \forall\, \alpha, \beta$. Thus $f_{02} + f_{20} \geq 0$ is sufficient for transformability (Theorem 4.10(i)). However, by Theorem 4.10(ii), it is also necessary if $\partial\aleph$ has at least 2 points. The special cases of empty boundary ($\aleph = \mathbb{C}$), empty region, or a single point boundary are omitted here.

(iii) Here, $\delta(\alpha, \beta) = \frac{1}{2}(f_{02} + f_{20}) + \frac{1}{4}(3f_{30} + f_{12})(x + u) - \frac{1}{4}(3f_{03} + f_{21})(y + v)$, where $\alpha = (x,y)$ and $\beta = (u,v)$. Using Theorem 4.10(i), the result follows.

Finally, for bounded regions in the complex plane, we can characterize transformability by its boundary.

Theorem 4.11 Let $\aleph$, given by (4.1) - (4.2), be bounded. Then $\aleph$ is P-transformable, if and only if all $\alpha,\beta \in \partial\aleph$, $\alpha \neq \beta \Rightarrow \delta(\alpha,\beta) \geq 0$.

Proof Necessity holds for any region, by Theorem 4.10(ii). To prove sufficiency, note that if $\aleph$ is bounded, there exist $\alpha_0, \beta_0 \in \mathrm{cl}(\aleph)$, such that $\sup_{\alpha,\beta\in\bar{\aleph}} \mathrm{Re}[\phi(\alpha, \bar{\beta})] = \mathrm{Re}[\phi(\alpha_0, \bar{\beta}_0)]$.
If $\alpha_0, \beta_0\, \partial\aleph$, then by hypothesis $\mathrm{Re}[\phi(\alpha_0, \bar{\beta}_0)] \leq 0$, and thus $\mathrm{Re}[\phi(\alpha, \bar{\beta})] \leq 0$ $\forall\ \alpha, \beta \in \mathrm{cl}(\aleph)$, which means transformability. If either α_0 or β_0 are not in $\partial\aleph$, say $\alpha_0 \notin \partial\aleph$, then writing $\aleph$ as the union of its connected components $\aleph_i$, that is $\aleph = \cup\, \aleph_i$, $i \in I$, we see that $\alpha_0 \in \aleph_{io}$ for some $i_0 \in I$. But $\aleph_{io}$ is a bounded domain on which $\alpha \rightarrow \mathrm{Re}[\phi(\alpha, \bar{\beta}_0)]$ has a maximum (at $\alpha = \alpha_0$). Now, recall that according to the maximum principle, if Ω is a domain and $h(\cdot)$ is analytic in Ω, then $\mathrm{Re}(h)$ does not have a local maximum in Ω, unless $h(\cdot)$ is constant. Thus $\alpha \rightarrow \phi(\alpha, \bar{\beta}_0)$ is constant in $\aleph_{io}$, and since it is a polynomial, it is constant everywhere. In particular, we have

$$0 \geq \phi(\beta_0, \bar{\beta}_0) = \phi(\alpha_0, \bar{\beta}_0) = \mathrm{Re}[\phi(\alpha_0, \bar{\beta}_0)]$$

and again transformability is satisfied. This completes the proof.

4.3 R-TRANSFORMABILITY

Corollary 4.10 shows that important regions are P-transformable in the sense of Definition 1. However, part (ii) of that corollary reveals that at least one simple region is excluded. Take, for example, the exterior to the unit disk:

$$\aleph = \{(x+iy) \colon 1-x^2-y^2 < 0\}\ . \tag{4.17}$$

Here $f_{02} + f_{20} = -2$, and according to Corollary 4.10(ii), this region is not transformable. It is indeed a striking fact. To better understand the situation, note that

$$\phi(\alpha, \bar{\beta}) = 1 - \alpha\bar{\beta}\ . \tag{4.18}$$

On the other hand, the unit disk

$$\aleph = \{(x+iy) \colon -1+x^2+y^2 < 0\} \tag{4.19}$$

generates

$$\phi(\alpha, \bar{\beta}) = -\ 1 + \alpha\bar{\beta}\ . \tag{4.20}$$

Figure 4.1 shows the map $\phi(\aleph)$ for both regions.

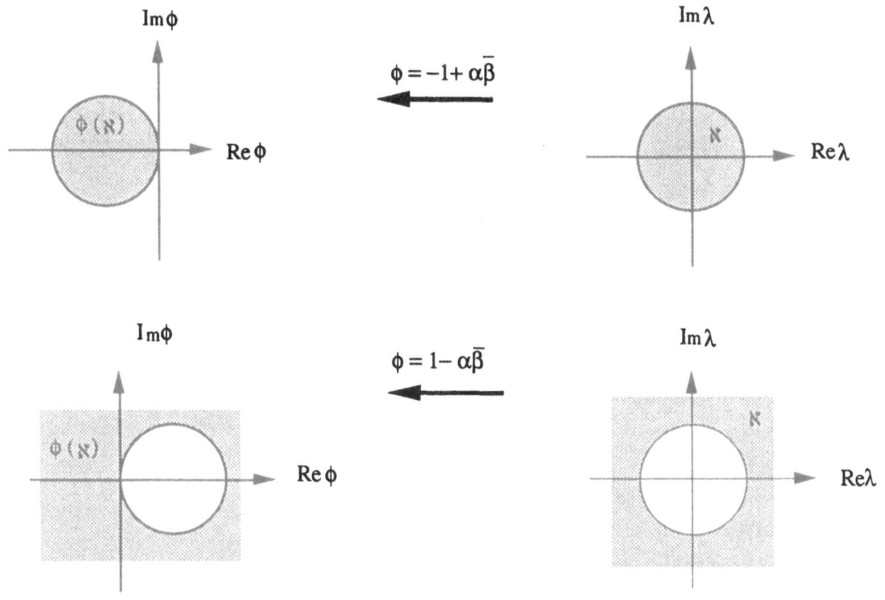

Figure 4.1 The map $\phi(\aleph)$.

The figure suggests that a slight change in (4.18) can make (4.17) transformable. We replace (4.18) by

$$\eta(\alpha, \bar{\beta}) = \frac{1-\alpha\bar{\beta}}{\alpha\bar{\beta}} = -1 + \frac{1}{\alpha\bar{\beta}} \,. \tag{4.21}$$

Figure 4.2 shows the dramatic change in the map.

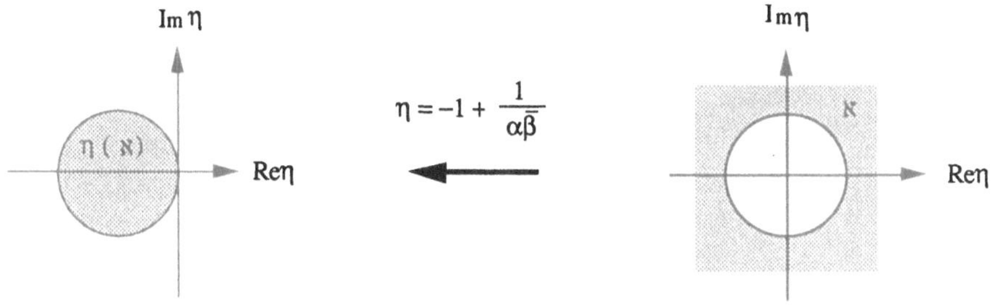

Figure 4.2 The map $\eta(\aleph)$, based on (4.21)

At the same time, (4.21) has another important property

$$\eta(\lambda, \bar{\lambda}) = \frac{1-\lambda\bar{\lambda}}{\lambda\bar{\lambda}} = \frac{1-x^2-y^2}{x^2+y^2} \,.$$

Since $x^2+y^2 > 0$ in $\aleph$, it follows that (4.21) maintains our basic relation (4.5); namely,

$$\aleph = \{\lambda \in C : \phi(\lambda,\bar{\lambda}) < 0\} .$$

Observing the proof of Theorem 4.3, we see that the rational function $\phi(\cdot)$ defined in (4.21) can replace $\phi(\cdot)$ in the theorem. We conclude that it is possible to extend the family of transformable regions by using a ***rational*** map of the form

$$\eta(\alpha, \bar{\beta}) = \frac{\phi(\alpha, \bar{\beta})}{\phi_1(\alpha, \bar{\beta})} . \tag{4.22}$$

We will specify the requirement on $\phi_1(\cdot)$ in a moment. Before doing so let us concentrate on an important family of regions.

In Section 3.3 we found that $\phi(\cdot)$ can be written as

$$\phi(\alpha, \bar{\beta}) = \sum_i \delta_i \psi_i(\alpha) \overline{\psi_i(\beta)} \tag{4.23}$$

where $\psi_i(\cdot)$ are certain polynomials, and δ_i are the eigenvalues of the matrix Φ defined in (4.6). The reader can verify that if Rank(Φ) - Sign(Φ)=2, then Φ has a single negative eigenvalue. Thus, for this case, absorbing δ_i by ψ_i, we have

$$\phi(\alpha, \bar{\beta}) = -\psi_1(\alpha)\overline{\psi_1(\beta)} + \sum_{i=2} \psi_i(\alpha) \overline{\psi_i(\beta)} \tag{4.24}$$

and

$$\text{(i)} \quad \phi(\lambda, \bar{\lambda}) = -|\psi_1(\lambda)|^2 + \sum_{i=2} |\psi_i(\lambda)|^2$$

$$\text{(ii)} \quad \aleph = \{\lambda \in C : \phi(\lambda,\bar{\lambda}) < 0\} . \tag{4.25}$$

Similarly to (4.21), we define

$$\eta(\alpha, \bar{\beta}) = \frac{\phi(\alpha, \bar{\beta})}{\psi_1(\alpha)\overline{\psi_1(\beta)}}$$

or

$$\eta(\alpha, \bar{\beta}) = -1 + \sum_{i=2} \frac{\psi_i(\alpha)\overline{\psi_i(\beta)}}{\psi_1(\alpha)\overline{\psi_1(\beta)}} . \tag{4.26}$$

First, note that

$$\eta(\lambda, \bar{\lambda}) < 0 \Leftrightarrow \phi(\lambda, \bar{\lambda}) < 0 . \tag{4.27}$$

Next, we wish to check transformability. If $\alpha,\beta \in \aleph$, then according to (4.25)

$$\sum_{i=2} |\psi_i(\alpha)|^2 < |\psi_1(\alpha)|^2$$

$$\sum_{i=2} |\psi_i(\beta)|^2 < |\psi_1(\beta)|^2 .$$

Thus,

$$|\psi_1(\alpha)\overline{\psi_1(\beta)}| = |\psi_1(\alpha)||\psi_1(\beta)| > (\sum_{i=2} |\psi_i(\alpha)|^2)^{\frac{1}{2}} (\sum_{i=2} |\psi_i(\beta)|^2)^{\frac{1}{2}} .$$

However, according to the Schwarz inequality, if $\{a_k\}$ and $\{b_k\}$ are complex numbers, then

$$\left|\sum_{k=1}^{n} a_k b_k\right|^2 \leq (\sum_{k=1}^{n} |a_k|^2)(\sum_{k=1}^{n} |b_k|^2)$$

so, in our case

$$\alpha, \beta \in \aleph \Rightarrow |\psi_1(\alpha)\overline{\psi_1(\beta)}| > \left|\sum_{i=2} \psi_i(\alpha)\overline{\psi_i(\beta)}\right| . \tag{4.28}$$

Combining (4.28) and (4.26), we see that

$$\alpha, \beta \in \aleph \Rightarrow |1 + \eta(\alpha, \bar{\beta})| < 1 . \tag{4.29}$$

This means that $\eta(\cdot)$ given by (4.26) maps $\alpha,\beta \in \aleph$ into a shifted unit disk contained in the LHP, and clearly transformability is satisfied. The relation (4.29) is depicted in Figure 4.3. The similarity to Figure 4.2 is evident.

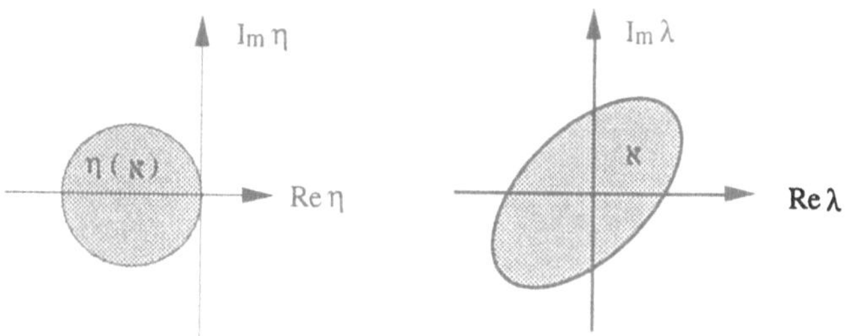

Figure 4.3 The map $\eta(\aleph)$ for Rank(Φ) - Sign (Φ) = 2.

We summarize our results in the following:

Theorem 4.12 A region $\aleph$, satisfying Rank(Φ) - Sign(Φ) = 2, is admissible. In particular, Theorems 4.3, 4.4, 4.6 and 4.7 hold, with $\eta(\cdot)$ (given by (4.26)) replacing $\phi(\cdot)$.

Now we return to the rational function $\eta(\cdot)$ given in (4.22). Recall that in order to apply Theorem 4.3, we need some requirements on $\eta(\alpha, \bar{\beta})$. First of all, we wish $\eta(\cdot)$ to be Hermitian so that $q(\cdot)$ is a real polynomial. Thus $\phi_1(.)$ must be hermitian. Second, at this stage, we wish to maintain (4.5); that is, $\aleph$ is uniquely characterized by $\phi(\lambda, \lambda) < 0$. Since we use $\eta(\cdot)$, we require that (4.27) holds. This is satisfied, if $\phi_1(\lambda,\lambda)$ is nonnegative and it is positive in $\aleph$. Collecting our requirements, we write :

$$\begin{aligned}
&\text{(i)} \quad \phi_1(\alpha, \bar{\beta}) \text{ is Hermitian,} \\
&\text{(ii)} \quad \phi_1(\lambda, \bar{\lambda}) \geq 0 \quad \forall\ \lambda \in C, \qquad\qquad (4.30)\\
&\text{(iii)} \quad \phi_1(\alpha, \bar{\beta}) \neq 0 \quad \forall\ \alpha, \beta \in \aleph .
\end{aligned}$$

Definition 4.4 A region $\aleph$ defined in (4.1) is R-transformable, if for $\eta(\cdot)$ defined in (4.22) and (4. 30), all $\alpha, \beta \in \aleph \Rightarrow \mathrm{Re}[\eta(\alpha, \bar{\beta})] < 0$.

Theorem 4.13 Suppose $\aleph$ is R-transformable. Then $\aleph$ is admissible. In particular, Theorems 4.3, 4.4, 4.6 and 4.7 hold, with $\eta(\cdot)$ replacing $\phi(\cdot)$.

Example 4.1.
According to Corollary 4.10(i), the left sector $\{(x+iy): -hx+y < 0\}$ is P-transformable. According to part (ii), the double sector $\{(x+iy):-h^2x^2+y^2 < 0\}$ is P-transformable, if and only if $h \leq 1$. According to Example 3.4, if $h > 1$ there exists no root clustering criterion, thus the double sector with $h > 1$ is not transformable (in any sense). Consider now the ***left sector*** $\aleph = \{(x+iy): -h^2x^2+y^2 < 0, x < 0; h > 1\}$. This region is not P-transformable . However, as we will see in a moment, it is R-transformable . Recall that for the double sector

$$\phi(\alpha, \beta) = -h^2\left(\frac{\alpha+\beta}{2}\right)^2 + \left(\frac{\alpha-\beta}{2i}\right)^2 = -(1+h^2)(\alpha^2+\beta^2) + 2(1-h^2)\alpha\beta$$

$$= -(1+h^2)(\alpha+\beta)^2 + 4\alpha\beta .$$

Choose

$$\phi_1(\alpha, \beta) = (1+h^2)(\alpha+\beta)^2 .$$

Note that this function satisfies (4.30). Thus, (4.22) takes the form

$$\eta(\alpha, \beta) = -1 + 2(1-H)\frac{\alpha\beta}{(\alpha+\beta)^2}$$

where

$$H = \frac{h^2 - 1}{h^2 + 1} \text{ and } h > 1 \quad \Rightarrow 0 < H < 1.$$

We now show that

$$\alpha, \beta \in \aleph \Rightarrow 2(1-H)\left|\frac{\alpha\beta}{(\alpha+\beta)^2}\right| < 1. \quad \text{Then, } \mathrm{Re}[\eta] < 0.$$

Use $\alpha = R_1 e^{j\theta_1}$, $\beta = R_2 e^{j\theta_2}$. Then

$$2(1-H)\left|\frac{\alpha\beta}{(\alpha+\beta)^2}\right| \le 2(1-H)\frac{|\alpha\beta|}{(\mathrm{Re}(\alpha+\beta))^2} = 2(1-H)\frac{R_1 R_2}{(R_1 \cos\theta_1 + R_2 \cos\theta_2)^2}.$$

It is sufficient to show that in $\aleph$

$$2(1-H)R_1R_2 < (R_1\cos\theta_1 + R_2\cos\theta_2)^2$$

or

$$R_1^2\cos^2\theta_1 + 2R_1R_2(\cos\theta_1\cos\theta_2 + H - 1) + R_2\cos^2\theta_2 > 0$$

and in quadratic form

$$[R_1\ R_2]\begin{bmatrix}\cos^2\theta_1 & \cos\theta_1\cos\theta_2 + H - 1 \\ \cos\theta_1\cos\theta_2 + H - 1 & \cos^2\theta_2\end{bmatrix}\begin{bmatrix}R_1 \\ R_2\end{bmatrix} > 0.$$

This is satisfied, provided

$$\cos^2\theta_1\cos^2\theta_2 > (\cos\theta_1\cos\theta_2 + H - 1)^2$$

or, provided

$$(H-1)^2 + 2(H-1)\cos\theta_1\cos\theta_2 < 0.$$

Since $H-1 < 0$, it is left to show that in $\aleph$ $2\cos\theta_1\cos\theta_2 - 1 > -H$.

Now, $\alpha \in \aleph \Leftrightarrow \dfrac{\alpha^2 + \bar{\alpha}^2}{2|\alpha|^2} > -H$, or $\cos 2\theta_1 > -H$, and $\cos\theta_1 < 0$.

Since $\cos 2\theta_1 = 2\cos^2\theta_1 - 1$, we have

$$\alpha \in \aleph \Leftrightarrow \cos^2\theta_1 > \frac{1-H}{2}, \quad \cos\theta_1 < 0$$

$$\Leftrightarrow |\cos\theta_1| > \sqrt{\frac{1-H}{2}}, \; \cos\theta_1 < 0$$

$$\Leftrightarrow |\cos\theta_1| < -\sqrt{\frac{1-H}{2}}.$$

Likewise

$$\beta \in \aleph \Leftrightarrow \cos\theta_2 < -\sqrt{\frac{1-H}{2}}\ .$$

Thus,
$2\cos\theta_1\cos\theta_2 - 1 = 2|\cos\theta_1||\cos\theta_2| - 1 > \frac{1-H}{2} - 1 = -H$, as required We conclude $\alpha,\beta \in \aleph \Rightarrow \mathrm{Re}[\eta] < 0$.

Example 4.2
Similarly to Example 4.1, it is possible to show that the ***left hyperbola***

$$\aleph = \{(x+iy) : 1-a^2x^2+b^2y^2 < 0\,,\ x < 0\,;\ (a/b)^2 > 1\}$$

is R-transformable. Here, we choose

$$\phi_1(\alpha, \beta) = \frac{a^2+b^2}{4}\,(\alpha+\beta)^2\,.$$

Remarks

1. The computation in Example 4.1 reveals that $|\eta+1| < 1$. This result holds in Example 4.2 as well. It means that in both examples, $\eta(\cdot)$ maps $\aleph$ (*left* sector and hyperbola) into the shifted unit disk as in Figure 4.3, although Rank (Φ) - Sign$(\Phi) \neq 2$.
2. A close look at Examples 4.1 and 4.2 reveals that the choice $\phi_1(\alpha,\beta) = -(\alpha+\beta)$ yields $\alpha,\beta \in \aleph \Rightarrow \mathrm{Re}[\eta] < 0$. However, (4.30)(ii) is not satisfied.

   In closing this section we mention that in Section 6.5 we present a more general subclass of R-transformability.

4.4 IR-TRANSFORMABILITY

In this section we wish to broaden our approach to its most general form. In Section 4.1 we start with a ***polynomial*** function $\mu(\cdot)$. Then, in Section 4.3, we relax the class of polynomials and allow a

rational function $\eta(\cdot)$. It is now left to consider an ***irrational*** function. In order to save space, we will not prove the equivalent of Theorem 4.3. Rather, we will construct all the necessary tools for such a proof. First, notice that in (4.22) $\phi_1\eta - \phi = 0$. This is a first order polynomial in η. Now, construct the polynomial

$$g(\eta;\alpha,\bar{\beta}) = \sum_{k=0}^{L} \phi_k(\alpha,\bar{\beta})\eta^k \tag{4.31}$$

such that

$$\begin{aligned}
&\text{(i)} && \phi_k(\alpha,\bar{\beta}),\ k = 0,1,\ldots,L, \text{ are Hermitian;} \\
&\text{(ii)} && \phi_L(\lambda,\bar{\lambda}) \geq 0 \quad \forall \quad \lambda \in C; \\
&\text{(iii)} && \phi_L(\alpha,\bar{\beta}) \neq 0 \quad \forall \quad \alpha,\beta \in \aleph; \\
&\text{(iv)} && \text{The roots of } g(\eta;\lambda,\bar{\lambda}) \text{ are all real } \forall\, \lambda \in C.
\end{aligned} \tag{4.32}$$

Define

$$\aleph = \{\lambda \in C : \phi_k(\lambda,\bar{\lambda}) > 0\,, \quad k = 0,1,\ldots,L\}. \tag{4.33}$$

Remarks

1. We can guarantee (4.32iv) using an aperiodicity criterion.
2. (4.33) implies that all the roots of $g(\eta;\lambda,\bar{\lambda})$ are negative $\forall\, \lambda \in \aleph$.

The importance of (4.33) becomes clear from the fact that $\aleph$ is constructed as an intersection of regions of the form (4.1) - (4.2). In fact, (4.33) is equivalent to :

$$\aleph = \bigcap_{k=0}^{L} \aleph_k\,, \quad \aleph_k = \{\lambda \in C : \phi_k(\lambda,\bar{\lambda}) > 0\}. \tag{4.34}$$

However, we do not define transformability, as in Definition 4.1, for each $\aleph_k$. Rather, we adopt the following.

Definition 4.5 A region $\aleph$ defined in (4.33), or equivalently (4.34), is ***IR- transformable***, if $g(\eta;\alpha,\bar{\beta})$ given by (4.31) - (4.32) is Hurwitz $\forall\, \alpha,\beta \in \aleph$.

Although for $g(\eta)$ to be Hurwitz we need L inequalities, in the important case where $\aleph$ is ***connected***, we need only a ***single inequality***.

Theorem 4.14 Let $\aleph$ in (4.33) be connected and such that (i) - (iv) in (4.32) are satisfied. Then, $g(\eta;\alpha,\bar{\beta})$ is Hurwitz for all $\alpha,\beta \in \aleph$, if and only if $\mathrm{Res}_\eta[g(\eta,\alpha,\bar{\beta}), g(-\eta;\beta,\bar{\alpha})] > 0$ for all $\alpha,\beta \in \aleph$, where, with respect to η, $\mathrm{Res}_\eta(\cdot)$ is the resultant of the two polynomials in $(\cdot)$, see Sec. 5.1.

Proof To simplify notations, write $g(\eta) = g(\eta;\alpha,\bar{\beta})$ where we understand that $g(\eta)$ is a complex polynomial in η with coefficients varying continuously with α and β. From Orlando's formula,

$\mathrm{Res}[g(\eta)], \bar{g}(-\eta)] > 0$ is the critical constraint of $g(\eta)$ with respect to the left half plane. This means that if for specific values of parameters $g(\eta)$ is Hurwitz, then $\mathrm{Res}[g(\eta), \bar{g}(-\eta)] > 0$ among other inequalities. Now, if we vary α and β in $\aleph$ continuously, this resultant vanishes as soon as a root of $g(\eta)$ hits the imaginary axis. However, for $\alpha = \beta \in \aleph$, $g(\eta)$ is Hurwitz since it is aperiodic with positive coefficients. Thus, as we vary α and β in $\aleph$, as long as the above resultant is positive, $g(\eta)$ is Hurwitz. This completes the proof.

Note that it is possible to relax (4.32) - (4.33) in the following way:

(i) $\phi_k(\alpha,\bar{\beta})$, $k = 0,1,\ldots,L$, are Hermitian,

(ii) $\phi_L(\alpha,\bar{\beta}) \neq 0 \quad \forall \quad \alpha,\beta \in \aleph$, (4.32)'

(iii) $\phi_L(\lambda,\bar{\lambda}) \geq 0 \quad \forall \quad \lambda \in C$

$$\aleph = \{\lambda \in C : g(\eta;\lambda,\bar{\lambda}) \text{ is aperiodic and stable}\} . \tag{4.33'}$$

According to this version, $\aleph$ combines the previous inequalities $\phi_k(\lambda,\bar{\lambda}) > 0$ and the aperiodicity inequalities.

Now we can state the following result.

Theorem 4.15 Let $\aleph \in C$, defined by (4.31) - (4.33), be IR-transformable. Let $\Delta(\lambda) = \sum_{i=0}^{n} a_i\lambda^i$ be a complex polynomial with roots $\{\lambda_i\}$. Let $q(\eta) = \sum_{i=0}^{n^2L} q_i\eta^i$, $q_{n^2L} > 0$, be the real polynomial with roots $\{\eta_{ij}\}$, where η_{ij} represents the roots of the polynomial $\sum_{k=0}^{L} \phi_k(\lambda_i, \bar{\lambda}_j)\eta^k$. Then, $\sigma(\Delta) \subset \aleph$, if and only if $q_i > 0$, $i = 0,. 1,\ldots, n^2L$.

Proof Sufficiency — $q_i > 0 \ \forall \ i$ implies that $q(\eta)$ has only negative real roots. In particular, since $g(\eta; \lambda, \bar{\lambda})$ has only real roots, all of them are negative. Thus $\sum\phi_k(\lambda, \bar{\lambda})\eta^k = \prod_k(\eta + \eta_k)$ with $\eta_k > 0$. we get $\phi_k(\lambda, \bar{\lambda}) > 0$, or $\sigma(\Delta) \subset \aleph$.

Necessity — if $\sigma(\Delta) \subset \aleph$ and $\aleph$ is IR-transformable, then $g(\eta;\lambda_i,\lambda_j)$ is Hurwitz, so all the roots η_{ij} of $g(\eta)$ have negative real parts. This implies (as in the proof of Theorem 4.3) that $q_i > 0 \ \forall \ i$.

The purpose of this theorem is to point out the ***structure*** of a general root clustering criterion. As mentioned before, the criterion is not ready yet for computation. We need a way to generate $q(\eta)$ from $\Delta(\lambda)$ and $\aleph$. Before doing so we present a few examples.

4.5 EXAMPLES

This section is devoted to a comparison of P-transformability and Rank(Φ)-Sign(Φ)= 2. The following examples show that neither one of them contains the other.

Example 4.3

Let a third order region be given by

$$\aleph = \{(x+iy) : 1-x(x^2+y^2) < 0\} \tag{4.35}$$

as shown in Figure 4.4.

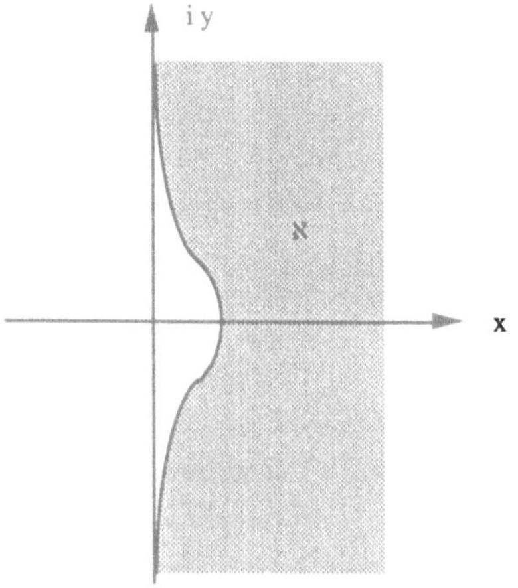

Figure 4.4 ℵ defined by (4.35)

We first show that this region is not P-transformable. In the proof of Theorem 4.10, we wrote an expression for $\delta(\alpha,\beta)$. If we use it in part (ii) of the theorem, we find that a necessary condition for transformability is

$$f_{20} + f_{02} + (f_{12} + 3f_{30})x - (f_{21}+3f_{03})y \geq 0 \tag{4.36}$$

for all $x + iy \in \aleph\backslash I$.

In this example, $\delta(\alpha,\alpha) = -4x$; thus, $\delta(1,1) < 0$. But $\alpha=1$ is a point of the boundary, and it is not isolated since $f(\alpha) = 0$, $f'(\alpha) = [-3 \ \ 0] \neq 0$. Thus (4.36) is not satisfied.

Next, we show that our region satisfies Rank(Φ)-Sign (Φ) = 2.

Since $\phi(\alpha, \bar{\beta}) = 1 - \frac{1}{2}\alpha^2\bar{\beta} - \frac{1}{2}\alpha\bar{\beta}^2$, it follows that

$$\Phi = \begin{bmatrix} 1 & 0 & 0 \\ 0 & 0 & -\frac{1}{2} \\ 0 & -\frac{1}{2} & 0 \end{bmatrix}.$$

This matrix has only one negative eigenvalue.

Example 4.4
Consider the fourth order region

$$\aleph = \{(x+iy) : -1 + ax^4 + by^2 < 0; \quad a,b > 0\} \tag{4.37}$$

which is shown in Figure 4.5.

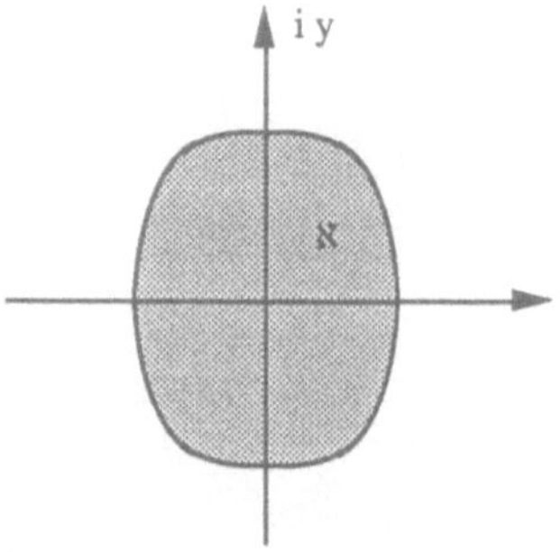

Figure 4.5 ℵ defined by (4.37)

We first show that this region does not satisfy Rank(Φ) - Sign(Φ) = 2. Expanding

$$\phi(\alpha, \bar{\beta}) = -1 + a\left(\frac{\alpha + \bar{\beta}}{2}\right)^4 + b\left(\frac{\alpha - \bar{\beta}}{2i}\right)^2, \quad \text{we find}$$

$$\Phi = \begin{bmatrix} -1 & 0 & -b/4 & 0 & a/16 \\ 0 & b/2 & 0 & a/4 & 0 \\ -b/4 & 0 & 3a/8 & 0 & 0 \\ 0 & a/4 & 0 & 0 & 0 \\ a/16 & 0 & 0 & 0 & 0 \end{bmatrix}.$$

The characteristic polynomial of Φ has the form

$$\Delta(\Phi) = \left(\lambda^2 - \frac{b}{2}\lambda - \frac{a^2}{16}\right)\left(\lambda^3 + \ldots + 6\left(\frac{a}{16}\right)^3\right)$$

Thus, the five real eigenvalues satisfy

$$\lambda_1\lambda_2 = -\frac{a^2}{16} < 0$$

and

$$\lambda_3\lambda_4\lambda_5 = -6\left(\frac{a}{16}\right)^3 < 0.$$

This shows that Φ has at least two negative eigenvalues. Next, we find conditions for P-transformability. First note that $\aleph$ in nonempty, bounded, connected, and algebraic. Thus, we can use Theorem 4.11 directly. To verify that $\forall\ \alpha,\beta \in \partial\aleph,\ \alpha \neq \beta \Rightarrow \delta(\alpha,\beta) \geq 0$, we can minimize $\delta(\alpha,\beta)$ over the compact curve $\partial\aleph$. Here, we just find sufficient conditions for $\delta(\alpha, \beta) \geq 0\ \forall\ \alpha,\ \beta \in \mathrm{cl}(\aleph)$ and use Theorem 4.10. Take $\alpha = x + iy$, $\beta = u + iv$ and use (4.15). Then,

$$\delta(\alpha, \beta) = \phi_{11} + \phi_{22} | S_2|^2 + 2\ \mathrm{Re}(\phi_{21}S_2 + \phi_{31}S_3)$$

$$= \frac{a}{8}\,(7x^2 + 10xu + 7u^2) + \frac{b}{2} - \frac{a}{8}\,(y + v)^2 .$$

Since the first term is always nonnegative (try!), $\delta(\alpha, \beta \geq 0$ in $\bar{\aleph}$ if

$$\frac{b}{2} - \frac{a}{8}\,(y + v)^2 \geq 0 \qquad \text{in } \bar{\aleph} .$$

Since in $\bar{\aleph}$, $|y| \leq \frac{1}{\sqrt{b}}$, $|v| \leq \frac{1}{\sqrt{b}}$, it follows that $\delta(\alpha, \beta) \geq 0$ in $\bar{\aleph}$ if

$$\frac{b}{2} - \frac{a}{8}\left(\frac{1}{\sqrt{b}} + \frac{1}{\sqrt{b}}\right)^2 \geq 0$$

or

$$b^2 - a \geq 0 .$$

We conclude that $\aleph$ in (4.37) is P-transformable if $b^2 \geq a$.

Example 4.5

Consider the fourth order "ellipse"

$$\aleph = \{(x+iy) : -1+ax^4 + by^4 < 0; \quad a,b > 0\} \tag{4.38}$$

as shown in Figure 4.6.

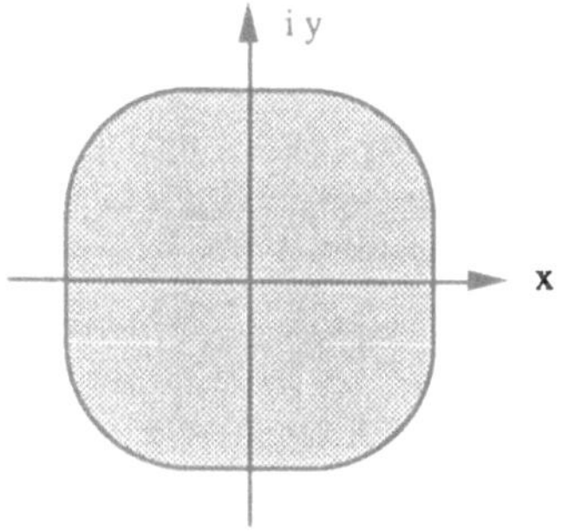

Figure 4.6 ℵ defined by (4.38)

As in Example 4.4, ℵ in (4.38) is nonempty, bounded, connected and algebraic. Expanding

$$\phi(\alpha, \beta) = -1 + a\left(\frac{\alpha+\beta}{2}\right)^4 + b\left(\frac{\alpha-\beta}{2i}\right)^4,$$

we find

$$\Phi = \begin{bmatrix} -1 & 0 & 0 & 0 & \frac{1}{16}(a+b) \\ 0 & 0 & 0 & \frac{1}{4}(a-b) & 0 \\ 0 & 0 & \frac{3}{8}(a+b) & 0 & 0 \\ 0 & \frac{1}{4}(a-b) & 0 & 0 & 0 \\ \frac{1}{16}(a+b) & 0 & 0 & 0 & 0 \end{bmatrix}$$

and using (4.15)

$$\delta(\alpha, \beta) = \tfrac{3}{8}(a+b)|\alpha+\beta|^2 + \tfrac{1}{2}(a-b)\mathrm{Re}(\alpha^2+\alpha\beta+\beta^2).$$

We see that if a=b, then $\delta(\alpha,\beta) \geq 0$. On the other hand, taking $\alpha = -\beta = x+iy$ in $\delta(\alpha,\beta)$, we obtain

$$\delta(\alpha, -\alpha) = \tfrac{1}{2}(a-b)(x^2-y^2).$$

But (x^2-y^2) is positive for $\alpha = a^{-1/4}$ and negative for $\alpha = ib^{-1/4}$ and in both cases α and $-\alpha$ belong to the boundary. Thus, $\delta(\alpha,\beta)$ takes negative values on $\partial\aleph$ if $a \neq b$. According to the necessary condition Theorem 4.10(ii), ℵ is not P-transformable if $a \neq b$. Next, we compute the characteristic polynomial of Φ:

$$\Delta(\Phi) = \left(\lambda - \tfrac{3}{8}(a+b)\right)\left(\lambda^2 - \left(\frac{a-b}{4}\right)^2\right)\left(\lambda^2 + \lambda - \left(\frac{a+b}{16}\right)^2\right)$$

and we see that this polynomial has one negative root, if and only if a=b. We conclude: $\aleph$ is both P-transformable and Rank (Φ)-Sign (Φ) = 2, if and only if a = b.

Example 4.6

Consider the fourth order "parabola"

$$\aleph = \{(x+iy) : x+y^4 < 0\} \tag{4.39}$$

which is shown in Figure 4.7.

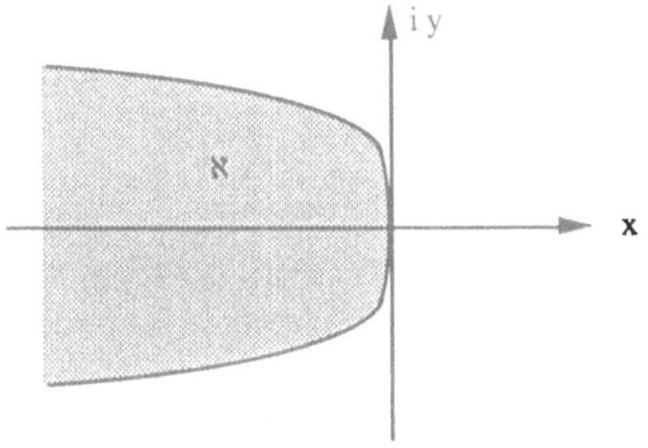

Figure 4.7 $\aleph$ defined by (4.39).

This region is connected, unbounded and algebraic. Note that $\aleph$ is a rational region, that is, its boundary is a rational curve. Indeed,

$$\partial\aleph = \{-t^4+it : t \in R\} .$$

Now, expanding

$$\phi(\alpha, \beta) = \tfrac{1}{2}(\alpha + \beta) + \left(\frac{\alpha - \beta}{2i}\right)^4$$

we find

$$\Phi = \begin{bmatrix} 0 & \frac{1}{2} & 0 & 0 & \frac{1}{16} \\ \frac{1}{2} & 0 & 0 & -\frac{1}{4} & 0 \\ 0 & 0 & \frac{3}{8} & 0 & 0 \\ 0 & -\frac{1}{4} & 0 & 0 & 0 \\ \frac{1}{16} & 0 & 0 & 0 & 0 \end{bmatrix}$$

The eigenvalues $\lambda_1,\lambda_2,\ldots,\lambda_5$ of Φ satisfy

$$\prod_{i=1}^{5}\lambda_i = |\Phi| = 3/8^5 > 0 .$$

Thus, Φ cannot have exactly one negative eigenvalue. Using (4.15), we have

$$\delta(\alpha,\beta) = \tfrac{3}{8}|S_2|^2 - \tfrac{1}{4}S_1\bar{S}_3 - \tfrac{1}{4}S_3\bar{S}_1$$

$$= \tfrac{3}{8}|\alpha+\beta|^2 - \tfrac{1}{2}\mathrm{Re}\left(\alpha^2+\alpha\beta+\beta^2\right).$$

Taking $\beta=0$ and $\alpha=-t^4+it$, we obtain

$$\delta(-t^4+it,0) = \tfrac{1}{8}t^8 + \tfrac{7}{8}t^2;$$

This clearly shows that $\delta(\alpha,\beta)$ takes negative values in $\partial\aleph$ and so the necessary condition Theorem 4.10, (ii) is not satisfied. We conclude that $\aleph$ is neither P-transformable nor Rank(Φ)-Sign(Φ) = 2.

An important consequence of the previous examples is :

$$\begin{aligned}&\text{(i)}\quad P-\text{transformability} \not\subset \text{Rank}(\Phi)-\text{Sign}(\Phi)=2\\ &\text{(ii)}\quad \text{Rank}(\Phi)-\text{Sign}(\Phi)=2 \not\subset P-\text{transformability}\end{aligned} \qquad (4.40)$$

Thus, we need both families of regions.

Let us continue by presenting some more examples.

Example 4.7

Consider the Cissoid of Diocles

$$\aleph = \{(x+iy) : x^3+y^2(x+a) < 0\} \qquad (4.41)$$

as shown in Figure 4.8, where $a > 0$.

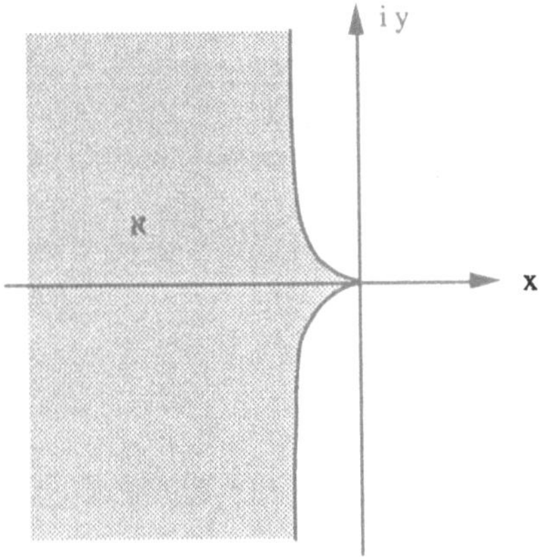

Figure 4.8 ℵ defined by (4.41).

Expanding

$$\phi(\alpha, \beta) = \left(\frac{\alpha + \beta}{2}\right)^3 + \left(\frac{\alpha - \beta}{2i}\right)^2 \left(\frac{\alpha + \beta}{2} + a\right),$$

we find

$$\Phi = \begin{bmatrix} 0 & 0 & -a \\ 0 & 2a & 2 \\ -a & 2 & 0 \end{bmatrix}.$$

The characteristic polynomial of Φ is

$$|\lambda I - \Phi| = \lambda^3 - 2a\lambda^2 - (4-a^2)\lambda + 2a^3.$$

Clearly, $-\lambda_1\lambda_2\lambda_3 = 2a^3 > 0$. Thus $\lambda_1\lambda_2\lambda_3 < 0$, and Φ has either a single negative (real) root or three such roots. However, if the latter holds, $-(\lambda_1+\lambda_2+\lambda_3) = -2a < 0$, implies $\lambda_1+\lambda_2+\lambda_3 > 0$, which is impossible. Thus, Φ has a single negative (real) root, and Rank(Φ)-Sign(Φ) = 2 .

Example 4.8

Consider the Lemniscate of Bernoulli

$$\aleph = \{(x+iy) : (x^2+y^2)^2 - a^2(x^2-y^2) < 0\} \tag{4.42}$$

as shown in figure 4.9.Expanding

$$\phi(\alpha,\beta) = \left[\left(\frac{\alpha+\beta}{2}\right)^2 + \left(\frac{\alpha-\beta}{2i}\right)^2\right]^2 - a^2\left[\left(\frac{\alpha+\beta}{2}\right)^2 - \left(\frac{\alpha-\beta}{2i}\right)^2\right],$$

we find

$$\Phi = \begin{bmatrix} 0 & 0 & -\frac{1}{2}a^2 \\ 0 & 0 & 0 \\ -\frac{1}{2}a^2 & 0 & 1 \end{bmatrix}$$

and

$$|\lambda I - \Phi| = \lambda(\lambda^2 - \lambda - \tfrac{1}{4}a^4).$$

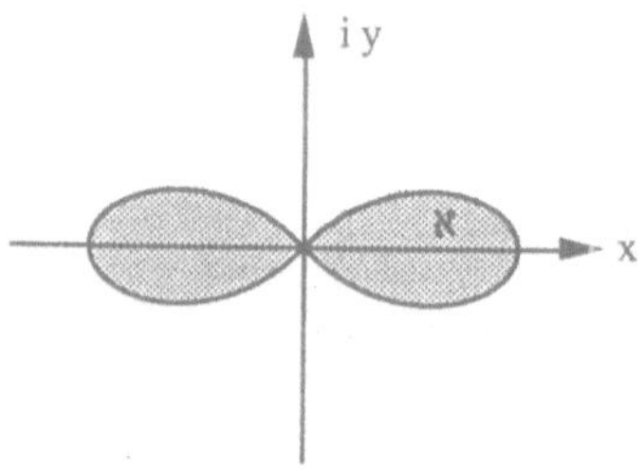

Figure 4.9 ℵ defined by (4.42).

Thus Φ has a single negative (real) root, and Rank(Φ)-Sign(Φ) = 2. In fact, Rank(Φ) = 2, and Sign(Φ) = 0.

NOTES AND REFERENCES

P-transformability was first defined by Gutman [1] and [2] and was later discussed by Gutman and Jury [1]. Theorem 4.2 is due to Chojnowski [1] and Gutman and Chojnowski [1]. Theorem 4.3 was discussed in the above papers. However, the complex version required special treatment. Later, it was observed that real and complex cases have the same structures. The fact that in Theorem 4.4 one needs ℵ to be simple is due to Chojnowski [1]. The original result, as well as Theorem 4.6, is due to Gutman [1]. Theorems 4.8 and 4.9 are developed in Araposthathis and Jury [1]. In Gutman [3], an attempt to develop a simple test for P-transformability, turned out to be false. Section 4.2 is due to Chojnowski [1] and can be found in Gutman and Chojnowski [1]. R and IR-transformability are basically due to Gutman [4], and further refined by Gutman and Chojnowski [1]. Examples 4.1 and 4.2 were specially constructed by Taub [1]. Finally, we mention that Routh [1] himself stated Theorem 4.3 for the left half plane, and is quoted in an important article by Fuller [1]. Surprisingly, the first result in this direction was published by Waring [1] as early as 1763.

Chapter 5 : ROOT CLUSTERING CRITERIA

In this chapter we describe root clustering criteria based on composite matrices and polynomials. We generate criteria for regions satisfying P-, R- or IR-transformability.

5.1 COMPOSITE POLYNOMIALS

Recall the basic root clustering inclusion

$$\sigma(A) \subset \aleph$$

where

$$\Delta(\lambda) = \sum_{i=0}^{n} a_i \lambda^i ,$$

and

$$\aleph = \{(x+iy) : f(x,y) < 0\} .$$

According to Theorem 4. 3, we need a polynomial of $q(\mu) = \sum_{i=0}^{n^2} q_i \mu^i$ whose roots are

$$\phi(\lambda_i, \bar{\lambda}_j) = f\left(\frac{\lambda_i + \bar{\lambda}_j}{2}, \frac{\lambda_i - \bar{\lambda}_j}{2i}\right).$$

Then, $\sigma(\Delta) \subset \aleph$ if and only if $q_i > 0 \ \forall\ i$.

In Theorem 4.13, the two variable polynomial $\phi(\lambda_i, \lambda_j)$ is replaced by the two variable rational function $\eta(\lambda_i, \lambda_j) = \dfrac{\phi(\lambda_i, \lambda_j)}{\phi_1(\lambda_i, \lambda_j)}$. In this section we construct the polynomial $q(\mu)$. However, we open with a broader question. Consider the complex polynomials

$$\text{(i)}\quad a(\lambda) = \sum_{i=0}^{n} a_i \lambda^i , \qquad a_n \neq 0$$

$$\text{(ii)}\quad b(s) = \sum_{i=0}^{m} b_i s^i , \qquad b_m \neq 0 \tag{5.1}$$

with roots $\{\lambda_i\}$ and $\{s_i\}$, respectively.

We are looking for the nm-th order composite polynomial $q(\eta)$ with the following nm roots

$$\eta_{ij} = \frac{\phi(\lambda_i, s_j)}{\phi_1(\lambda_i, s_j)} \tag{5.2}$$

where

$$\text{(i)}\quad \phi(\lambda_i, s_j) = \sum_{p,q} c_{pq}\lambda_i^p s_j^q$$

$$\text{(ii)}\quad \phi_1(\lambda_i, s_j) = \sum_{t,u} d_{tu}\lambda_i^t s_j^u . \tag{5.3}$$

Here, c_{pq} and d_{tu} are given coefficients, and $\phi_1(\cdot)$ does not vanish identically. By definition, $q(\eta)$ is obtained by eliminating λ and s from the following set of equations

$$a(\lambda)=0,\ \ b(s) = 0,\ \ \phi_1(\lambda,s)\eta - \phi(\lambda,s) = 0 . \tag{5.4}$$

Toward this end we need the following concept.

Definition 5.1. A **resultant** $R(u, v)$ of polynomials $u(\lambda) = \sum_{i=0}^{n} u_i\lambda^i$ and $v(\lambda) = \sum_{i=0}^{m} v_i\lambda^i$ is a scalar which is nonzero if and only if $u(\lambda)$ and $v(\lambda)$ are relatively prime.

In other words, $R(u,v) = 0$ if and only if $u(\lambda)$ and $v(\lambda)$ have at least one common root. One form for $R(u,v)$ is

$$R(u, v) \triangleq u_n^m v_m^n \prod_{i=1}^{n}\prod_{j=1}^{m}(\lambda_i - s_j) \tag{5.5}$$

where $\{\lambda_i\}$ and $\{s_j\}$ are the roots of $u(\lambda)$ and $v(\lambda)$, respectively. Since

$$u(\lambda) = u_n(\lambda - \lambda_1)(\lambda - \lambda_2)\ldots(\lambda - \lambda_n) = u_n\prod_{i=1}^{n}(\lambda - \lambda_i)$$

and

$$v(\lambda) = v_m(\lambda - s_1)(\lambda - s_2)\ldots(\lambda - s_m) = v_m\prod_{j=1}^{m}(\lambda - s_j)$$

we have

$$R(u, v) = u_n^m \prod_{i=1}^{n} v(\lambda_i),$$

$$R(u, v) = (-1)^{nm} v_m^n \prod_{j=1}^{m} u(s_j).$$

According to Theorem 3.4, if $\sigma(A) = \{\lambda_i\}$, the eigenvalues of $v(A)$ are $\{v(\lambda_i)\}$. Since $|A| = \prod_{i=1}^{n}\lambda_i$, we have $|v(A)| = \prod_{i=1}^{n} v(\lambda_i)$ and, for the usual situation, $u_n = 1$, we may conclude the following result.

Theorem 5.1 The polynomials $u(\lambda)$ and $v(\lambda)$ have a common factor, if and only if $|v(A)| = |u(B)| = 0$. In fact $R(u,v) = |v(A)|$, where A and B have characteristic polynomials $u(\lambda)$ and $v(\lambda)$, respectively.

Next, we introduce the Sylvester matrix $S(u,v)$

$$S(u,v) = \begin{bmatrix} u_n & u_{n-1} & \cdots & u_0 & 0 & \cdots & 0 \\ 0 & u_n & \cdots & u_1 & u_0 & \cdots & 0 \\ \vdots & \vdots & & & & & \vdots \\ 0 & 0 & \cdots & u_n & \cdots & & u_0 \\ \hline v_m & v_{m-1} & \cdots & v_0 & \cdots & & 0 \\ 0 & v_m & \cdots & v_1 & v_0 & \cdots & 0 \\ \vdots & \vdots & & & & & \vdots \\ 0 & 0 & \cdots & v_m & \cdots & & v_0 \end{bmatrix} \begin{matrix} \} \text{ m rows} \\ \\ \} \text{ n rows} \end{matrix} \tag{5.6}$$

Theorem 5.2 The polynomials $u(\lambda)$ and $v(\lambda)$ have a common factor, if and only if $|S(u,v)| = 0$. In fact $R(u,v) = |S(u,v)|$.

Proof Necessity — if $u(x)$ and $v(x)$ have a common factor $r(x)$ with $\deg(r) \geq 1$, then $u(x) = u_1(x)\, r(x)$, and $v(x) = v_1(x)\, r(x)$. Multiply $u = u_1 r$ by v and $v = v_1 r$ by u. Then we have $u_1 rv = v_1 r$, or $uv_1 = vu_1$, which can be written $uv_1 + v\tilde{u}_1 = 0$, where $\tilde{u}_1 = -u_1$. In particular,

$$\left(\sum_{i=0}^{n} u_i x^i\right)\left(\sum_{i=0}^{m-1} d_i x^i\right) + \left(\sum_{i=0}^{m} v_i x^i\right)\left(\sum_{i=0}^{n} c_i x^i\right) = 0 . \tag{5.7}$$

Writing this equation in terms of x and equating its coefficients to zero, we obtain

$$[d_{m-1}\ d_{m-2} \cdots d_0\ \vdots\ c_{n-1}\ c_{n-2} \cdots c_0]\ S(u,v) = 0 .$$

For this set of equations to have a non-trivial solution, it is necessary and sufficient that $Res(u,v) = |S(u,v)| = 0$. Sufficiency — if $|S(u,v)| = 0$, we can repeat the above steps backwards, and deduce the existence of polynomials $u_1(x)$ and $v_1(x)$ such that $uv_1 = vu_1$, $\deg u_1 \leq n - 1$, $\deg v_1 \leq m-1$. However, since the coefficients $\{d_i\}$ and $\{c_i\}$ are chosen arbitrarily, we can choose them such that u_1 is a factor in u and v_1 is a factor in v. Thus, for instance, $v = v_1 h$. Multiplying $uv_1 = vu_1$ by h, we obtain $uv_1h = vu_1h$. Using $v = v_1h$, we find $uv = vu_1h$, so that $u = u_1h$, and we conclude that $h(x)$ is a common factor of $u(x)$ and $v(x)$. This completes the proof.

The first form of the resultant is connected with Bezout (1764). We associate with polynomials $u(\lambda)$ and $v(\lambda)$ the quotient

$$B(u, v) = \frac{u(\lambda)v(s) - u(s)v(\lambda)}{\lambda - s} .$$

(5.8)

Since the numerator has (λ-s) as a factor, B(u,v) is a polynomial and we may write

$$B(u, v) = \sum_{i=1}^{n} \sum_{j=1}^{n} z_{ij} \lambda^{i-1} s^{j-1}$$

(5.9)

where $u(\lambda)$ and $v(\lambda)$ are written as polynomials of the same degree n. Notice that the nxn matrix $Z = [z_{ij}]$ is ***symmetric*** .

Theorem 5.3 The polynomials $u(\lambda)$, $a_n = 1$, and $v(\lambda)$ have a common factor, if and only if $|Z| = 0$. In fact $Z = Tv(A)$, where A is the companion matrix of $u(\lambda)$ and

$$T = \begin{bmatrix} a_1 & a_2 \cdots & a_{n-1} & 1 \\ a_2 & a_3 & 1 & \\ \vdots & \vdots & & \\ a_{n-1} & 1 & & \bigcirc \\ 1 & & & \end{bmatrix} .$$

An alternative way of constructing Z is as follows. Write the Sylvester matrix S with m=n (if m≠n, put zero for the proper entries) and define

$$S = \begin{bmatrix} S_1 & S_2 \\ S_3 & S_4 \end{bmatrix} \qquad J = \begin{bmatrix} O & 1 \\ 1 & O \end{bmatrix} .$$

Then

$$Z = S_1 S_4 J - S_3 S_2 J . \tag{5.10}$$

Finally, we comment on the dimension of the resultant (Def. 5.1).

(i) In Theorem 5.1, Dim(R) = min(m,n),

(ii) In Theorem 5.2, Dim(R) = n+m,

(iii) In Theorem 5.3, Dim(R) = max(m,n).

We now return to the set of equations (5.4). If (5.4) holds, then $a(\lambda)$ and $\phi_1(\lambda,s)\eta+\phi(\lambda,s)$ have a common factor. Thus

$$q(\eta; s) \triangleq \text{Res}[a(\lambda), \quad \phi_1(\lambda, s)\eta + \phi(\lambda, s)] = 0$$

and

$$b(s) = 0 .$$

For these two equations to hold, we must have

$$q(\eta) \triangleq \mathrm{Res}[b(s), \quad q(\eta; s)] = 0 .$$

We now conclude

Theorem 5.4 The composite polynomial $q(\eta)$, defined by (5.1) - (5.3), is given by

(i) $q(\eta;s) = \mathrm{Res}[a(\lambda), \phi_1(\lambda,s)\eta - \phi(\lambda,s)]$
(ii) $q(\eta) = \mathrm{Res}[b(s), q(\eta;s)]$.

As an illustration, consider the left half plane and the polynomial $a(\lambda)$. Here $\phi(\lambda,s) = \lambda+s$, and $\phi_1(\lambda,s) = 1$. Thus, according to Theorems 5.1 and 5.4,

$$q(\eta; s) = \mathrm{Res}[a(\lambda), \eta - (\lambda + s)] = \mathrm{Res}[a(\lambda), \lambda - (\eta - s)]$$
$$= |a(B)| = a(\eta - s)$$
$$= \sum_{i=0}^{n} a_i(\eta - s)^i .$$

Thus,

$$q(\eta) = \mathrm{Res}[a(s), a(\eta-s)] . \tag{5.11}$$

Note that from root clustering point of view, this is equivalent to Orlando's

$$q(\eta) = \mathrm{Res}[a(\eta+s), a(\eta-s)] .$$

Next, consider the unit disk, where $\phi(\lambda,s) = -1+\lambda s$. According to Theorems 5.1 and 5.4,

$$q(\eta; s) = \mathrm{Res}[a(\lambda), \eta - (-1 + \lambda s)]$$
$$= \mathrm{Res}[a(\lambda), -s\lambda + (\eta + 1)]$$
$$= \mathrm{Res}[a(\lambda), \lambda - \frac{\eta + 1}{s}]$$
$$= a\left(\frac{\eta + 1}{s}\right) = \sum_{i=0}^{n} a_i(\eta + 1)^i s^{n-i} .$$

Thus,

$$q(\eta) = \mathrm{Res}[a(s), \sum_i a_i(\eta+1)^i s^{n-1}]$$
$$= \mathrm{Res}[a(s), a(\frac{\eta+1}{s})] . \qquad (5.12)$$

Note that from the root clustering point of view, this is equivalent to

$$q(\eta) = \mathrm{Res}[a((\eta+1)s), a(\frac{\eta+1}{s})] .$$

Example 5.1

Let

$a(\lambda) = \lambda^2+\lambda+1$

$b(s) = s^2+2s+5$

$\eta(\lambda,s) = \lambda+s$.

According to (5.10),

$$q(\eta) = \mathrm{Res}[b(s), a(\eta-s)]$$
$$= \mathrm{Res}[s^2+2s+5, s^2-(2\eta+1)s + (\eta^2+\eta+1)] .$$

Let

$$B = \begin{bmatrix} 0 & 1 \\ -5 & -2 \end{bmatrix} .$$

Then, according to Theorem 5.1,

$$q(\eta) = |B^2 - (2\eta - 1)B + (\eta^2 + \eta + 1)I|$$
$$= \begin{vmatrix} \eta^2 + \eta - 4 & -(2\eta + 3) \\ 5(2\eta + 3) & \eta^2 + 5\eta + 2 \end{vmatrix}$$
$$= \eta^4 + 6\eta^3 + 23\eta^2 + 42\eta + 37 .$$

Example 5.2

Let

$a(\lambda) = \lambda^2+5\lambda+6$

$b(s) = s^2+2s+2$

$\mu(\lambda,s) = -1+\lambda s$.

According to (5.11),

$$q(\eta) = \mathrm{Res}[s^2+2s+2,\ 6s^2+5(\eta+1)s + (\eta+1)^2] .$$

Let

$$B = \begin{bmatrix} 0 & 1 \\ -2 & -2 \end{bmatrix} .$$

Then, according to Theorem 5.1,

$$q(\eta) = |6B^2 + 5(\eta + 1)B + (\eta + 1)^2 I|$$

$$= \begin{vmatrix} \eta^2 + 2\eta - 11 & 5\eta - 7 \\ -2(5\eta - 7) & \eta^2 - 8\eta + 3 \end{vmatrix}$$

$$= \eta^4 + 6\eta^3 + 26\eta^2 - 46\eta + 65 \quad .$$

Example 5.3

$a(\lambda)$ and $b(s)$ as in Example 5.2, but

$$\eta(\lambda, s) = \frac{1 - \lambda s}{\lambda s} .$$

According to Theorems 5.1 and 5.4

$$q(\eta; s) = \mathrm{Res}[a(\lambda),\ \lambda s\eta + \lambda s - 1]$$

$$= \mathrm{Res}[a(\lambda),\ s(\eta + 1)\lambda - 1]$$

$$= \mathrm{Res}[a(\lambda),\ \lambda - \frac{1}{(\eta + 1)s}]$$

$$= a(\frac{1}{(\eta + 1)s}) = \sum a_i(\eta + 1)^{n-i}s^{n-i} \quad .$$

Then

$$q(\eta) = \mathrm{Res}[b(s),\ \sum_i a_i(\eta + 1)^{n-i}s^{n-i}]$$

$$= \mathrm{Res}[s^2 + 2s + 2,\quad 6(\eta + 1)^2 s^2 + 5(\eta + 1)s + 1]$$

$$= |6(\eta + 1)^2B^2 + 5(\eta + 1)B + I|$$

$$= \begin{vmatrix} -12(\eta+1)^2 + 1 & -(\eta+1)(12\eta+7) \\ 2(\eta+1)(12\eta+7) & 12(\eta+1)^2 - 10(\eta+1) + 1 \end{vmatrix}$$

$$= 144\eta^4 + 456\eta^3 + 578\eta^2 + 354\eta + 65 \ .$$

We are now ready to extend our results. Let the polynomials $a(\lambda)$ and $b(s)$ be given as in (5.1). In Theorem 4.14 we are looking for a polynomial $q(\eta)$ whose roots $\{\eta_{ij}\}$ are the roots of the polynomial $g(\eta; \lambda_i, s_j) = \sum_{k=0}^{L} \phi_k(\lambda_i, s_j)\eta^k$.

Observing Theorem 5.4, the reader can verify the following:

Corollary 5.4 The composite polynomial $q(\eta)$ defined in (5.1) and (4.31) is given by:

(i) $q(\eta; \lambda) = \mathrm{Res}[b(x), \sum_{k=0}^{L} \phi_k(\lambda, x)\eta^k]$

(ii) $q(\eta) = \mathrm{Res}[a(\lambda), q(\eta; \lambda)]$.

5.2 COMPOSITE MATRICES

In this section we develop a matrix version of composite polynomials. In particular, given two matrices $A \in C^{n\times n}$ and $B \in C^{m\times m}$, we are looking for a matrix $C \in C^{nm\times nm}$ whose eigenvalues are $\phi(\lambda_i,s_j)$, where $\phi(\cdot)$ is a given polynomial, and $\sigma(A) = \{\lambda_i\}$, $\sigma(B) = \{s_i\}$. Let us construct C for the case $n = m = 2$. If $x = [x_1 \ \ x_2]'$ and $y = [y_1 \ \ y_2]'$ are the eigenvectors of A and B, corresponding to λ_1 and s_1, respectively, we may write

$$\lambda_1 x_1 = a_{11} x_1 + a_{12} x_2 \qquad\qquad s_1 y_1 = b_{11} y_1 + b_{12} y_2$$

$$\lambda_1 x_2 = a_{21} x_1 + a_{22} x_2 \qquad\qquad s_1 y_2 = b_{21} y_1 + b_{22} y_2 .$$

Next, perform the following multiplications:

$$\begin{aligned}
\lambda_1 s_1 x_1 y_1 &= a_{11}b_{11}x_1y_1 + a_{11}b_{12}x_1y_2 + a_{12}b_{11}x_2y_1 + a_{12}b_{12}x_2y_2 \\
\lambda_1 s_1 x_1 y_2 &= a_{11}b_{21}x_1y_1 + a_{11}b_{22}x_1y_2 + a_{12}b_{21}x_2y_1 + a_{12}b_{22}x_2y_2 \\
\lambda_1 s_1 x_2 y_1 &= a_{21}b_{11}x_1 y_1 + a_{21}b_{12}x_1y_2 + a_{22}b_{11}x_2y_1 + a_{22}b_{12}x_2y_2 \\
\lambda_1 s_1 x_2 y_2 &= a_{21}b_{21}x_1 y_1 + a_{21}b_{22}x_1 y_2 + a_{22}b_{21}x_2y_1 + a_{22}b_{22}x_2y_2 .
\end{aligned} \qquad (5.13)$$

Define the vector

$$z = \begin{bmatrix} x_1 y_1 \\ x_1 y_2 \\ x_2 y_1 \\ x_2 y_2 \end{bmatrix} \qquad (5.14)$$

and the matrix

$$C = \begin{bmatrix} a_{11}b_{11} & a_{11}b_{12} & a_{12}b_{11} & a_{12}b_{12} \\ a_{11}b_{21} & a_{11}b_{22} & a_{12}b_{21} & a_{12}b_{22} \\ a_{21}b_{11} & a_{21}b_{12} & a_{22}b_{11} & a_{22}b_{12} \\ a_{21}b_{21} & a_{21}b_{22} & a_{22}b_{21} & a_{22}b_{22} \end{bmatrix} . \tag{5.15}$$

Thus, (5.13) takes the form

$$Cz = \lambda_1 s_1 z . \tag{5.16}$$

We see that z is an eigenvector of C, corresponding to the eigenvalue $\lambda_1 s_1$. Similarly, we can show that C has $\lambda_1 s_2$, $\lambda_2 s_2$ as eigenvalues. A close look at C reveals the special structure

$$C = \begin{bmatrix} a_{11}B & a_{12}B \\ a_{21}B & a_{22}B \end{bmatrix}, \quad \text{or } C = [a_{ij}B] . \tag{5.17}$$

Given $A \in C^{n\times n}$ and $B \in C^{n\times n}$, it is now natural to define a new product

$$A \otimes B = [a_{ij}B] , \tag{5.18}$$

called the ***Kronecker product*** . The dimension of $A \otimes B$ is $nm \times nm$. For $n = m = 2$, we have found that $C = A \otimes B$ has $\{\lambda_i \; s_j\}$ as eigenvalues. We will show in a moment that this holds for any dimension, but first we wish to present some useful properties of the Kronecker product.

Properties of the Kronecker Product

1. $(\mu A) \otimes B = A \otimes (\mu B) \;\; \forall \mu \in C;$
2. $(A+B) \otimes C = A \otimes C + B \otimes C;$
3. $A \otimes (B+C) = A \otimes B + A \otimes C;$
4. $A \otimes (B \otimes C) = (A \otimes B) \otimes C;$
5. $(A \otimes B)' = A' \otimes B';$
6. If $A,C \in C^{n\times n}$ and $B,D \in C^{n\times n}$, then
   $(A \otimes B)(C \otimes D) = AC \otimes BD;$
7. $\det(A \otimes B) = (\det A)^n (\det B)^m ;$
8. $\mathrm{tr}(A \otimes B) = (\mathrm{tr}A)(\mathrm{tr}B)$, where tr denotes trace;
9. $r(A \otimes B) = r(A)\, r(B)$, where r denotes rank;
10. $(A \otimes B)^{-1} = A^{-1} \otimes B^{-1}$, if A and B are nonsingular;
11. If $A_i \in C^{n\times n}$ and $B_j \in C^{m\times m}$, then
    $\prod_i (A_i \otimes B_i) = (\prod_i A_i) \otimes (\prod_j B_j)$.

In 1900, Stephanos proved a beautiful result.

Theorem 5.5 Consider $A \in C^{n\times n}$, $B \in C^{m\times m}$ with spectra $\sigma(A) = \{\lambda_i\}$ and $\sigma(B) = \{s_i\}$. The eigenvalues of the composite matrix $\phi(A \otimes B) \in C^{nm\times nm}$ given by

$$\phi(A \otimes B) = \sum_{p,q} \phi_{pq} A^p \otimes B^q \tag{5.19}$$

are the nm numbers

$$\phi(\lambda_i, s_j) = \sum_{p,q} \phi_{pq} \lambda_i^p s_j^q . \tag{5.20}$$

Proof Let $J_1 = P^{-1}AP$ and $J_2 = Q^{-1}BQ$ be the Jordan canonical forms of A and B, respectively. Since J_1 and J_2 are upper triangular with diagonals consisting of the eigenvalues, it follows that $J_1{}^p$ is an upper triangular matrix with $\lambda_1{}^p, \lambda_2{}^p,\ldots,\lambda_n{}^p$ along the main diagonal. A similar property holds for $J_2{}^q$. From the structure of the Kornecker product, it follows that $J_1{}^p \otimes J_2{}^p$ is also upper triangular with $\{\lambda_i{}^p \lambda_j{}^q\}$ along the main diagonal. Thus, $\phi(J_1 \otimes J_2)$ is an upper triangular matrix with $\{\phi(\lambda_i, \lambda_j)\}$ along the main diagonal. In other words, $\phi(J_1 \otimes J_2)$ has eigenvalues $\{\phi(\lambda_i, \lambda_j)\}$. To complete the proof we now show that $\phi(J_1 \otimes J_2)$ and $\phi(A \otimes B)$ have the same eigenvalues. Using property 11, we find

$$\begin{aligned} J_1^p \otimes J_2^q &= (P^{-1}AP)^p \otimes (Q^{-1}BQ)^q \\ &= P^{-1}A^pP \otimes Q^{-1}B^qQ \\ &= (P^{-1} \otimes Q^{-1})(A^p \otimes B^q)(P \otimes Q) . \end{aligned}$$

Using property 10,

$$J_1^p \otimes J_2^q = (P \otimes Q)^{-1}(A^p \otimes B^q)(P \otimes Q)$$

and hence

$$\phi(J_1 \otimes J_2) = (P \otimes Q)^{-1}\phi(A \otimes B)(P \otimes Q) . \tag{5.21}$$

This shows that $\phi(J_1 \otimes J_2)$ and $\phi(A\otimes B)$ are similar and so they have the same eigenvalues. This completes the proof.

By construction, Theorem 5.5 is equivalent to the following:

Corollary 5.5 The roots of the polynomial $q(\mu) = |\mu I - \phi(A \otimes B)|$ are $\phi(\lambda_i, s_j) = \sum_{p,q} \phi_{pq}\lambda_i^p s_j^q$.

Next, suppose that the polynomial matrix

$$g(\eta; A \otimes B) = \sum_{k=0}^{L} \phi_k(A \otimes B)\eta^k \tag{5.22}$$

is the composite matrix of the polynomial

$$g(\eta; \lambda, s) = \sum_{k=0}^{L} \phi_k(\lambda, s)\eta^k ; \tag{5.23}$$

that is,

$$\phi_k(\lambda, s) = \sum_{k=0} \phi_{pqk} \lambda^p s^q , \qquad \phi_k(A \otimes B) = \sum_{p,q} \phi_{pqk} A^p \otimes B^q . \tag{5.24}$$

Using arguments similar to those used in the proof of Theorem 5.5, we find that $g(\eta; A\otimes B)$ is similar to $g(\eta; J_1 \otimes J_2)$. Thus,

$$|g(\eta; A \otimes B)| = |g(\eta; J_1 \otimes J_2)| = \prod_{i,j} \sum_{k=0}^{L} \phi_k(\lambda_i, s_j)\eta^k . \tag{5.25}$$

We now have a matrix version of Theorem 5.3.

Theorem 5.6 Let $\sigma(A) = \{\lambda_i\}$, $\sigma(B) = \{s_j\}$. The polynomial $q(\eta)$ whose roots are those of $g(\eta;\lambda_i,s_j)$ is given by $q(\eta) = \det[g(\eta;A\otimes B)]$.

Let us now define a new product of two real matrices, $A,B \in R^{n\times n}$. The **bialterante product** of A and B, written $A\odot B$ is a real matrix of dimension $m = \frac{1}{2} n(n-1)$ with elements $A \odot B_{pq,rs}$, where

$$A \odot B_{pq,rs} = \frac{1}{2}\begin{vmatrix} a_{pr} & a_{ps} \\ b_{qr} & b_{qs} \end{vmatrix} + \frac{1}{2}\begin{vmatrix} b_{pr} & b_{ps} \\ a_{qr} & a_{qs} \end{vmatrix} \tag{5.26}$$

$$p = 2,3,\ldots,n \; ; \qquad q = 1,2,\ldots,p-1$$
$$r = 2,3,\ldots,n \; ; \qquad s = 1,2,\ldots,r-1 .$$

In particular, the bialternate product of A with itself is $A \odot A$, where

$$A \odot A_{pq,\, rs} = \begin{vmatrix} a_{pr} & a_{ps} \\ a_{qr} & a_{qs} \end{vmatrix}.$$

A powerful result due to Stephanos, and similar to Theorem 5.5, is the following:

Theorem 5.7 Let $A \in R^{n\times n}$, with spectrum $\sigma(A) = \{\lambda_i\}$. The eigenvalues of the composite matrix $\phi(A \odot A) \in R^{m\times m}$, given by

$$\phi(A \odot A) = \sum_{p,\,q} \phi_{pq}\, A^p \odot A^q,$$

where ϕ_{pq} are real numbers,

are the $m = \frac{1}{2}n(n-1)$ numbers

$$\theta(\lambda_i, \lambda_j) = \frac{1}{2} \sum_{p,\,q} \phi_{pq}(\lambda_i^p \lambda_j^q + \lambda_j^p \lambda_i^q)$$

$i = 2,3,\ldots,n$
$j = 1,2,\ldots,i\text{-}1.$

5.3 ROOT CLUSTERING CRITERIA

The criteria of this section are natural consequences of previous results. We will start with criteria for polynomials and then discuss matrices. We do so since, in the case where a polynomial is given, it is computationally simpler to follow the polynomial version. In what follows we adopt the notation "Coef $q(\mu)$" for "all the coefficients in the polynomial $q(\mu)$."
Note that by now we have three types of transformability:

(i) Polynomial (P) - Definition 4.1,
(ii) Rational (R) - Definition 4.4,
(iii) Irrational (IR) - Definition 4.5.

Also note that if $\aleph = \cap \aleph_i$ and each $\aleph_i$ is admissible (that is, has a criterion), then $\aleph$ is also admissible and its criterion is the union of the $\aleph_i$ -criteria.

Combining Theorems 4.3 and 5.4, we state

Theorem 5.8 Let $\aleph$ be P-transformable (Def. 4.1), and consider the complex polynomial $\Delta(\lambda) = \sum_{i=0}^{n} a_i\lambda^i$, $a_n > 0$. Then $\sigma(\Delta) \subset \aleph$ if and only if Coef $q(\eta) > 0$ where $q(\eta) = \mathrm{Res}[\Delta(\lambda),\ q(\eta;\lambda)]$, and $q(\eta;\lambda) = \mathrm{Res}[\bar{\Delta}(s),\ \eta - \phi(\lambda, s)]$.

Combining (4.12) and Theorem 4.6, and using the elimination system (5.4), we find

Corollary 5.8 Let $\aleph$ be symmetric and P–transformable, and let $\Delta(\lambda) = \sum_{i=0}^{n} a_i \lambda^i$, $a_n > 0$, be a real polynomial. $\sigma(\Delta) \subset \aleph$ if and only if

(i) Coef $\tilde{q}(\eta) > 0$, where $\tilde{q}(\eta) = \mathrm{Res}[\Delta(\lambda), \eta - \phi(\lambda, \lambda)]$

(ii) Coef $(q(\eta)/\tilde{q}(\eta))^{\frac{1}{2}} > 0$.

Example 5.4

Let

$\Delta(s) = s^2 + 2s + 2$

$\aleph = \{(x+iy): -16 + x^2 + 4y^2 < 0\}$.

We wish to test the truth of the inclusion $\sigma(\Delta) \subset \aleph$. By Corollary 4.10 (ii), all elipses are transformable. Thus, we may apply Theorem 5.4. Note that

$$\phi(\alpha, \beta) = -16 + \left(\frac{\alpha+\beta}{2}\right)^2 + 4\left(\frac{\alpha-\beta}{2i}\right)^2$$

$$= \tfrac{1}{4}(-64 - 3\alpha^2 - 3\beta^2 + 10\alpha\beta) .$$

Since the factor $\frac{1}{4}$ does not influence the final result, we will omit it. It is left for the reader to verify that

$$q(\eta, \lambda) = \begin{vmatrix} 1 & 2 & 2 & 0 \\ 0 & 1 & 2 & 2 \\ 3 & -10\lambda & \eta + 64 + 3\lambda^2 & 0 \\ 0 & 3 & -10\lambda & \eta + 64 + 3\lambda^2 \end{vmatrix}$$

$$= 9\lambda^4 + 60\lambda^3 + (6\eta + 566)\lambda^2 + (20\eta + 1280)\lambda + (\eta + 64)(\eta + 58) + 36 .$$

Next,

$$q(\eta; A) = 9A^4 + 60A^3 + (6\eta + 566)A^2 + (20\eta + 1280)A + [(\eta + 64)(\eta + 58) + 36]I$$

where

$$A = \begin{bmatrix} 0 & 1 \\ -2 & -2 \end{bmatrix} .$$

Thus

$$q(\eta) = |q(\eta; A)| = \begin{vmatrix} \eta^2 + 110\eta + 2820 & 8\eta + 268 \\ -(16\eta + 536) & \eta^2 + 174\eta + 7404 \end{vmatrix}$$

$$= (\eta^2 + 110\eta + 2820)(\eta^2 + 174\eta + 7404) + (16\eta + 536)(8\eta + 268) .$$

Clearly, Coef $q(\eta) > 0$. Thus all the roots of $\Delta(\lambda)$ are clustered in the given ellipse.

Combining Theorems 4.4 and 5.4, we obtain a result which can be applied to the rest of the theorems.

Theorem 5.9 Let $\aleph$ be simple (Def. 3.1) and P-transformable (Def. 4.1), and consider $\Delta(\lambda) = \sum_{i=0}^{n} a_i\lambda^i, \quad a_n > 0.$ Then

$$\sigma(\Delta) \subset \mathrm{cl}(\aleph) \quad \text{if and only if} \quad \mathrm{Coef}\ q(\eta) \geq 0 ,$$

where

$$q(\eta) = \mathrm{Res}[\Delta(\lambda), q(\eta; \lambda)], \text{ and } q(\eta; \lambda) = \mathrm{Res}[\bar{\Delta}(s), \eta - \phi(\lambda, s)].$$

Combining Theorems 4.13 and 5.4 we have

Theorem 5.10 Let $\aleph$ be R – transformable and $\Delta(\lambda) = \sum^{n} a_i\lambda^i$, $a_n > 0$. Then

$$\sigma(\Delta) \subset \aleph \text{ if and only if } \mathrm{Coef}\ q(\eta) > 0$$

where

$$q(\eta) = \mathrm{Res}[\Delta(\lambda), q(\eta; \lambda)], \text{ and } q(\eta; \lambda) = \mathrm{Res}[\bar{\Delta}(s), \phi_1(\lambda, s)\eta - \phi(\lambda, s)].$$

A special case of this theorem is the following

Corollary 5.10 Suppose $\aleph$ satisfies Rank (Φ) – Sign $(\Phi) = 2$, and let $\Delta(\lambda) = \sum_{i=0}^{n} a_i\lambda^i$, $a_n > 0$. Let $q(\eta; \lambda) = \mathrm{Res}[\bar{\Delta}(s), \eta\psi_1(\lambda)\bar{\psi}_1(s) - \phi(\lambda, s)]$

$$= \mathrm{Res}[\bar{\Delta}(s), \eta\psi_1(\lambda)\psi_1(s) - \sum_{i=2} \psi_i(\lambda)\bar{\psi}_i(s)] .$$

Then, $\sigma(\Delta) \subset \aleph$ if and only if Coef $q(\eta) > 0$

where

$$q(\eta) = \mathrm{Res}[\Delta(\lambda), q(\eta;\lambda)].$$

The importance of Theorem 5.10 goes beyond what we might expect at first thought. Consider the ***double*** hyperbola

$$\aleph = \{(x+iy)\colon 1 - a^2 x^2 + b^2 y^2 < 0\} .$$

If $(a/b)^2 \leq 1$, then $\aleph$ is P-transformable and the root clustering criterion is similar to that of Example 5.4. If $(a/b)^2 > 1$, then according to Example 3.4, $\aleph$ is not transformable. However, we see from Example 4.2 that the ***left*** hyperbola

$$\aleph = \{(x+iy) : 1-a^2x^2+b^2 y^2 < 0\} \cap \{(x+iy) : x < 0\} \qquad (5.27)$$

is transformable with respect to the rational function $h = \phi/-\phi_1$. But $\eta(\lambda,\bar{\lambda}) < 0$ implies $\phi(\lambda,\bar{\lambda}) < 0$; that is, $1-a^2x^2 + b^2y^2 < 0$, which is the double hyperbola, not the left branch. If we go back to the proof of the root clustering theorem, we find that the necessity part remains unchanged. In the sufficiency part, however, we have to add a criterion with respect to the LHP. By so doing we restrict ourselves to the left hyperbola, without changing the necessity part. As a result, the criterion for the left hyperbola is the ***union of the criteria*** with respect to

(i) $\alpha + \beta$ (LHP),

(ii) $$\frac{-(a^2+b^2)(\alpha^2+\beta^2)-2(a^2-b^2)\alpha\beta+4}{(a^2+b^2)(\alpha+\beta)^2} \quad \text{(double hyperbola)}.$$

As a conclusion, we have

Theorem 5.11 Consider the left hyperbola (5. 27), with $(a/b)^2 > 1$. Let $\Delta(\lambda) \sum_{i=0}^{n} a_i\lambda^i$, $a_n > 0$, and $q(\eta;\lambda) = \text{Res}[\bar{\Delta}(s), (1+\eta)(a^2+b^2)s^2 + 2\lambda((a^2+b^2)\eta + a^2 - b^2)s + \lambda^2(a^2+b^2)(\eta+1)-4]$. Then all the roots of $\Delta(\lambda)$ lie in the left hyperbola (5.27), if and only if

(i) $\Delta(\lambda)$ is Hurwitz

(ii) Coef $q(\eta) > 0$, where $q(\eta) = \text{Res}[\Delta(\lambda), q(\eta;\lambda)]$.

In our opening remarks to this section, we mentioned that if two regions are admissible, their intersection is admissible as well. Now, with the aid of R-transformability, a new horizon is opened. Although the double hyperbola, with $(a/b)^2 > 1$, is not admissible, that of its intersection with the LHP is. Another comment is now in order. As a matter of fact, the denominator $\phi_1(\alpha,\beta)$ in (ii) can be taken as $-(a^2+b^2)(\alpha+\beta)$; that is, first order rather than second order. However, this $\phi_1(.)$ does not satisfy our requirement $\phi_1(\lambda,\bar{\lambda}) \geq 0$, $\forall \lambda \in C$. In the more general case (4.31), $\phi_1(\lambda,\bar{\lambda}) \geq 0$ simply means that the leading coefficient of $g(\cdot)$ is nonnegative.

Example 5.5

Let

$$\aleph = \{(x+iy) : 1-2x^2+y^2 < 0\} \cap \{(x+iy) : x < 0\}$$

and

$$\Delta(\lambda) = \lambda^2 + 4\lambda + 5 .$$

Then

$$q(\eta;\lambda) = \mathrm{Res}[x^2+4x+5,\ 3(\eta+1)x^2+2\lambda(3\eta+1)x+3\lambda^2(\eta+1)-4]$$

$$= \begin{vmatrix} 1 & 4 & 5 & 0 \\ 0 & 1 & 4 & 5 \\ 3(\eta+1) & 2\lambda(3\eta+1) & 3\lambda^2(\eta+1)-4 & 0 \\ 0 & 3(\eta+1) & 2\lambda(3\eta+1) & 3\lambda^2(\eta+1)-4 \end{vmatrix}$$

$$q(\eta) = \mathrm{Res}[\Delta(\lambda),\ q(\eta;\lambda)] =$$

$$= 8294400\eta^4 + 18690048\eta^3 + 15750144\eta^2 + 5879808\eta + 820224\,.$$

Thus, $q_i > 0$. Since $\Delta(\lambda)$ is Hurwitz, all the roots of $\Delta(\lambda)$ lie in the left hyperbola.

Finally, combining Theorems 4.15 and 5.4, we have

Theorem 5.12 Let $\aleph$ be IR−transformable, $\Delta(\lambda) = \sum_{i=0}^{n} a_i\lambda^i$, $a_n > 0$, and

$$q(\eta;\lambda) = \mathrm{Res}[\bar{\Delta}(s),\ \sum_{k=0}^{L} \phi_k(\lambda, s)\eta^k]\,.$$

Then, $\sigma(\Delta) \subset \aleph$ if and only if Coef $q(\eta) > 0$ where

$$q(\eta) = \mathrm{Res}[\Delta(\lambda),\ q(\eta;\lambda)]\,.$$

Consider, once more, the ***left*** hyperbola (4.66), with $(a/b)^2 > 1$. Recall that using the concept of R-transformability, $\eta(\lambda,\lambda) < 0$ is equivalent to the ***double*** hyperbola; thus, we have to add, artificially, the condition "$\Delta(\lambda)$ is Hurwitz". With the aid of IR-transformability, this is done automatically. Let $g(\cdot)$ in (4.31) be a second order polynomial in η

$$g(\eta;\alpha,\beta) = (\nu\eta-\mu)(\eta-\mu_2) \tag{5.28}$$

$$= \phi_2\eta^2 + \phi_1\eta + \phi_0\,.$$

In accordance with (5.28), we choose

(i) $\mu(\alpha,\beta) = -(a^2+b^2)(\alpha^2+\beta^2) - 2(a^2-b^2)\alpha\beta + 4$;

(ii) $\nu(\alpha,\beta) = (a^2+b^2)(\alpha+\beta)^2$;

(iii) $\mu_2(\alpha,\beta) = \alpha+\beta$.

We leave for the reader to prove that

(i) $g(\eta;\lambda,\bar{\lambda})$ is aperiodic $\forall\, \lambda \in \mathbb{C}$. (In fact, it holds for all $\{\nu, \mu, \eta_2\}$ in (5.28).)

(ii) (5.27) with $(a/b)^2 > 1$ is IR-transformable.

Now, we may apply Theorem 5.12

Example 5.6

We solve Example 5.5 using Theorem 5.12.

$$q(\eta;\lambda) = \text{Res}[x^2+4x+5,\ (\nu(\lambda,x)\eta-\mu(\lambda,x))(\eta-\mu_2(\lambda,x))]$$
$$= \text{Res}[x^2+4x+5,\ -3(1+\eta)x^3 + (3(\eta+1) - \lambda(9\eta+5))x^2 + (4+2\lambda(\eta-\lambda)(3\eta+1) - 3\lambda^2(\eta+1))x + (\eta-\lambda)(3\lambda^2(\eta+1)-4)]$$

$$= \begin{vmatrix} 1 & 4 & 5 & 0 & 0 \\ 0 & 1 & 4 & 5 & 0 \\ 0 & 0 & 1 & 4 & 5 \\ -3(\eta+1) & \begin{matrix}3(\eta+1)\\ -\lambda(9\eta+5)\end{matrix} & \begin{matrix}4+2\lambda(\eta-\lambda)(3\eta+1)\\ -3\lambda^2(\eta+1)\end{matrix} & (\eta-\lambda)(3\lambda^2(\eta+1)-4 & 0 \\ 0 & -3(\eta+1) & \begin{matrix}3(\eta+1)\\ -\lambda(9\eta+5)\end{matrix} & \begin{matrix}4+2\lambda(\eta-\lambda)(3\eta+1)\\ -3\lambda^2(\eta+1)\end{matrix} & (\eta-\lambda)(3\lambda^2(\eta+1)-4) \end{vmatrix}$$

Using a symbolic manipulation program, we find $q(\eta)$, and verify that $q_i > 0$.
We now turn to the matrix version. Combining Theorem 4.3 and Corollary 5.6, we obtain

Theorem 5.13 Let $\aleph$ be P-transformable, and consider $A \in \mathbb{C}^{n\times n}$. Let $\phi(A \otimes \bar{A}) = \sum_{i,j} \phi_{ij} A^i \otimes \bar{A}^j$, where the ϕ_{ij}'s are defined in (4.6). Then $\sigma(A) \subset \aleph$, if and only if $\text{Coef}\ |\mu I - \phi(A \otimes \bar{A})| > 0$.

Example 5.7

We solve Example 5.4 using Theorem 5.13. Here,

$$\phi(A\otimes A) = -64I - 3A^2 \otimes I - 3I \otimes A^2 + 10A \otimes A$$

where

$$A = \begin{bmatrix} 0 & 1 \\ -2 & -2 \end{bmatrix}.$$

Thus

$$\phi(A \otimes A) = \begin{bmatrix} -52 & 6 & 6 & 10 \\ -12 & -64 & -20 & -14 \\ -12 & -20 & -64 & -14 \\ 40 & 28 & 28 & -36 \end{bmatrix}$$

and

$$|\mu I - \phi| = \mu^4 + 216\mu^3 + 17360\mu^2 + 613888\mu + 8053760 .$$

Combining Theorems 4.4 and 5.6, we obtain a result which can be extended to the rest of the theorems.

Theorem 5.14 Let $\aleph$ be P-transformable, and consider $A \in C^{n\times n}$. Let $\phi(A \otimes \bar{A}) = \sum_{i,j} \phi_{ij} A^i \otimes \bar{A}^j$, where the ϕ_{ij}' s are defined in (4. 6). Then

$$\sigma(A) \subset cl(\aleph), \text{ if and only if Coef } |\mu I - \phi(A \otimes \bar{A})| \geq 0.$$

Combining Theorems 4.13 and 5.6, we obtain

Theorem 5.15 Let $\aleph$ be R-transformable, and consider $A \in C^{n\times n}$. Then

$$\sigma(A) \subset \aleph, \text{ if and only if Coef } |\phi_1(A \otimes \bar{A})\eta + \phi(A \otimes \bar{A})| > 0.$$

A special case of this theorem is the following.

Theorem 5.16 Suppose $\aleph$ satisfies Rank(ϕ) - Sign(ϕ) = 2, and let $A \in C^{n\times n}$. Then $\sigma(A) \subset \aleph$, if and only if Coef $|\psi_1(A) \otimes \overline{\psi_1(A)}\ \eta - \psi(A \otimes \bar{A})| > 0$.

Combining Theorems 4.15 and 5.6, we obtain

Theorem 5.17 Let $\aleph$ be IR-transformable, and consider $A \in C^{n\times n}$. Then $\sigma(A) \subset \aleph$, if and only if Coef $|\sum_{k=0}^{L} \phi_k(A \otimes \bar{A})\eta^k| > 0$, where $\phi_k(A \otimes \bar{A}) = \sum_{i,j} \phi_{ijk} A^i \otimes \bar{A}^j$.

In the case where $\aleph$ is symmetric, Theorems 4.5 and 5.8 imply that

$$\theta(\lambda_i, \lambda_j) = \frac{1}{2} \sum_{p,q} \phi_{pq}(\lambda_i^p \lambda_j^q + \lambda_j^p \lambda_i^q)$$

$$= \frac{1}{2}[\phi(\lambda_i, \lambda_j) + \phi(\lambda_j, \lambda_i)]$$

$$= \phi(\lambda_i, \lambda_j).$$

Thus, combining Theorems 4.6 and 5.7, we obtain

Theorem 5.18 Let $\aleph$ be symmetric and P-transformable, and consider $A \in R^{nxn}$.

Let $\phi(A \odot A) = \sum_{i,j} \phi_{ij} A^i \odot A^j$, where the ϕ_{ij}' s are defined in (4. 6). Then $\sigma(A) \subset \aleph$, if and only if

(i) $\quad \text{Coef} \mid \lambda I - \sum_i f_{io} A^i \mid > 0$,

(ii) $\quad \text{Coef} \mid \mu I - \phi(A \odot A) \mid > 0$.

Comparing Theorems 5.15 and 5.18, we see that in the former, the number of inequalities is n^2, while in the latter, part (i) has n and part (ii) has $\frac{1}{2} n (n - 1)$ inequalities.

Theorems 4.8 and 5.18 imply

Corollary 5.18. The eigenvalues of a real matrix A are clustered in the open left half plane, if and only if

(i) $(-1)^n |A| > 0$,

(ii) $\text{Coef} |\mu I - A \odot I - I \odot A| > 0$.

A similar result holds for the unit disk.

Next, Theorem 5.18 is directly extended to R- and IR-transformable regions.

Theorem 5.19 Let $\aleph$ be symmetric and IR-transformable, and consider $A \in R^{nxn}$.

Let $\phi_k(\lambda) = \phi_k(\lambda, \lambda)$ where $\phi_k(\lambda, s)$ is given by (5. 23), $\phi_k(A \odot A) = \sum_{i,j} \phi_{ijk} A^i \odot A^j$, and recall Section 4.4. Then $\sigma(A) \subset \aleph$, if and only if

(i) $\quad \text{Coef} \mid \sum_{k=0}^{L} \phi_k(A) \eta^k \mid > 0$,

(ii) $\quad \text{Coef} \mid \sum_{k=0}^{L} \phi_k(A \odot A) \eta^k \mid > 0$.

Corollary 5.19 If $\aleph$ contains the entire real line, then only condition (ii) is required.

5.4 SOME IMPORTANT REGIONS

After presenting a general theory for root clustering, we focus on three important regions: the left hyperbola, the left sector and aperiodicity. We devote a special section to these regions, since they play an important role in the theory of dynamical systems. As we will see, the criteria are closely related to our previous results.

THE LEFT HYPERBOLA

First, recall Theorem 3.8 . That result is concerned with root clustering (or distribution) of a given ***polynomial*** with respect to the hyperbola (and other conic sections). Here, we are concerned with ***matrix*** root clustering with respect to the left hyperbola

$$\aleph = \{(x+iy): 1-a^2x^2 + b^2y^2 < 0\} \cap ((x+iy) : x < 0\} \tag{5.29}$$

where the first member of $\aleph$ is the double hyperbola, and the second is the left half plane. For the double hyperbola

$$\phi(\alpha, \beta) = 1 - a^2\left(\frac{\alpha+\beta}{2}\right)^2 + b^2\left(\frac{\alpha+\beta}{2i}\right)^2$$

$$= 1 - \frac{a^2+b^2}{4}(\alpha^2+\beta^2) + \frac{b^2-a^2}{2}\alpha\beta .$$

Now, if $(a/b)^2 \leq 1$, the double hyperbola is P-transformable and we can state the following theorem.

Theorem 5.20 Let $\aleph$ be the left hyperbola (5.29), with $(a/b)^2 \leq 1$, and consider $A \in C^{n\times n}$. Then, $\sigma(A) \subset \aleph$ if and only if

(i) A is Hurwitz,

(ii) $\text{Coef}\,|\,\mu I - \phi(A \otimes \bar{A})| = \text{Coef}\,|(\mu - 1)I + \frac{a^2+b^2}{4}(A^2 \otimes I + I \otimes \bar{A}^2)$

$$- \frac{b^2-a^2}{2} A \otimes \bar{A}| > 0.$$

In the case where A is a real matrix, we can use $\phi(A \odot A)$, and condition (i) is properly modified to take care of the real eigenvalues.

Corollary 5.20 Let $\aleph$ be the left hyperbola (5.29), with $(a/b)^2 \leq 1$, and consider $A \in R^{n\times n}$. Then, $\sigma(A) \subset \aleph$, if and only if

(i) $A + \frac{1}{|a|} I$ is Hurwitz ,

(ii) $\text{Coef}\,|\,\mu I - \phi(A \odot \bar{A})| > 0.$

Now, we discuss the case $(a/b)^2 > 1$. Based on Remark 2 following Example 4.2, we have

Theorem 5.21 Let $\aleph$ be the left hyperbola (5.29), with $(a/b)^2 > 1$, and consider $A \in C^{nxn}$. Then, $\sigma(A) \subset \aleph$, if and only if

(i) A is Hurwitz,

(ii) $\mathrm{Coef}\,|(A \otimes I + I \otimes \bar{A})\eta + \phi(A \otimes \bar{A})| > 0.$

In case A is a real matrix, we have

Corollary 5.21 Let $\aleph$ be the left hyperbola (5.29), with $(a/b)^2 > 1$ and consider $A \in R^{nxn}$. Then $\sigma(A) \subset \aleph$, if and only if

(i) $A + \frac{1}{|a|} I$ is Hurwitz,

(ii) $\mathrm{Coef}\,|(A \odot I + I \odot A)\eta + \phi(A \odot A)| > 0.$

THE LEFT SECTOR

Consider the left sector shown in Figure 5.1. For simplicity we suppose that A is a ***real*** matrix. Thus,

$$\sigma(A) \subset \text{left sector} \Leftrightarrow \sigma(A) \subset \{(x+iy): \delta x - y < 0\}. \qquad (5.30)$$

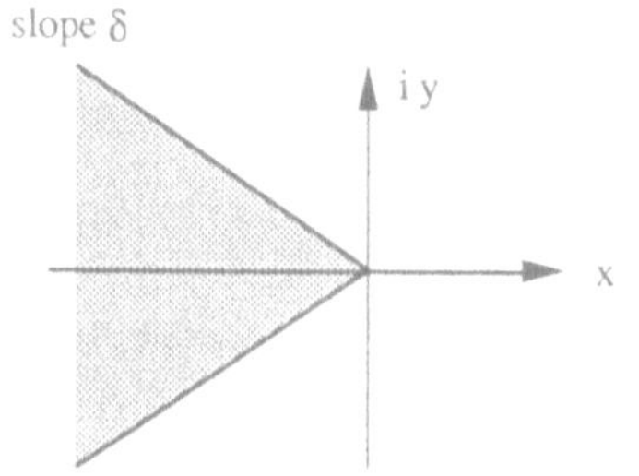

Figure 5.1 The left sector

The right hand side of (5.30) implies

$$\phi(\alpha, \beta) = (\delta + j)\alpha + (\delta - j)\beta, \qquad j = \sqrt{-1}.$$

Since every first-order region is P-transformable, we can apply Theorem 5.13 directly, to obtain:

Theorem 5.22 Let $\aleph$ be the left sector of Figure 5.1, and consider $A \in R^{n\times n}$. Then

$$\sigma(A) \subset \aleph \Leftrightarrow \text{Coef } |\eta I - (\delta+j)A \otimes I - (\delta-j)I \otimes A| > 0.$$

We now present an alternative method. Let $\sigma(A) = \{\lambda_i\}$. Since the eigenvalues of

$$B = \begin{bmatrix} \cos\alpha & -\sin\alpha \\ \sin\alpha & \cos\alpha \end{bmatrix}$$

are $\cos\alpha \pm j\sin\alpha$, it follows that the eigenvalues of $A_1 = B \otimes A$ are $\lambda_i (\cos\alpha \pm j \sin\alpha)$. As a result, $\sigma(A)$ is in the left sector, if and only if $\sigma(A_1)$ is in the open left half plane. Dividing B by the positive number $\sin\alpha$, we may replace it by:

$$B = \begin{bmatrix} \delta & -1 \\ 1 & \delta \end{bmatrix}$$

without changing our conclusion. Thus we have

Theorem 5.23 Let $\aleph$ be the left sector, and consider $A \in R^{n\times n}$. Then $\sigma(A) \subset \aleph$, if and only if the characteristic polynomial $c(\eta) = |\eta I - A_1|$, where

$$A_1 = \begin{bmatrix} A\delta & -A \\ A & A\delta \end{bmatrix}$$

is Hurwitz.

To simplify the theorem, note that

$$|sI-B| = s^2 - 2\delta s + (\delta^2+1)$$

and let

$$\Delta(\lambda) = |\lambda I - A| = \sum_{i=0}^{n} a_i \lambda^i .$$

Applying Theorem 5.3 to $\mu = \lambda s$, we obtain

$$q(\eta;\lambda) = \mathrm{Res}[x^2 - 2\delta x + \delta^2 + 1\,,\ \eta - \lambda x]$$

$$= \begin{vmatrix} 1 & -2\delta & \delta^2+1 \\ -\lambda & \eta & 0 \\ 0 & -\lambda & \eta \end{vmatrix}$$

$$= (\delta^2+1)\lambda^2 - 2\delta\eta\lambda + \eta^2 \ .$$

Thus, the polynomial whose roots are $\lambda_i(\cos\alpha \pm j\sin\alpha)$, is given by

$$c(\eta) = |\eta^2 I - 2\delta A\eta + (\delta^2+1)A^2|$$

where A is the companion matrix of $\Delta(\lambda)$, but more naturally, the original matrix A. We conclude

Theorem 5.24 Let $\aleph$ be the left sector, and consider $A \in R^{n\times n}$. Then $\sigma(A) \subset \aleph$ if and only if the polynomial

$$c(\eta) = |\,\eta^2 I - 2\delta A\eta + (\delta^2+1)A^2\,|$$

is Hurwitz.

Example 5.8

Given $\Delta(\lambda) = \lambda^2 + 2\lambda + 2$, we wish to test $\sigma(\,\Delta\,) \subset \aleph$, where $\aleph$ is the open left sector with unit slope $(\delta = 1)$. Using

$$A = \begin{bmatrix} 0 & 1 \\ -2 & -2 \end{bmatrix}$$

in Theorem 5.24, we find

$$c(\eta) = \begin{vmatrix} \eta^2 - 4 & -2(\eta+2) \\ 4(\eta+2) & \eta^2 + 4\eta + 4 \end{vmatrix}$$

$$= \eta^4 + 4\eta^3 + 8\eta^2 + 16\eta$$

$$= (\eta^3 + 4\eta^2 + 8\eta + 16)\eta \ .$$

Applying Routh table to the third order polynomial $\eta^3 + 4\eta^2 + 8\eta + 16$, we find

1	8
1	4
4	
4	.

We conclude that $c(\eta)$ is marginally stable, with one root at the origin. As a consequence, $\Delta(\lambda)$ does not have all its roots in the open left sector. Since $c(\eta)$ is marginally stable, $\Delta(\lambda)$ has all its roots in the closed left sector. Indeed, the two roots of $\Delta(\lambda)$ are located on the boundary.

APERIODICITY

A real matrix is said to be ***aperiodic*** if all its eigenvalues are real; thus, the solution of the vector equation $\dot{x} = Ax$ contains only pure exponential functions. Note that according to Corollary 4.10, (ii), the real axis

$$\aleph = \{(x+iy) : y^2 \leq 0\} \tag{5.31}$$

is P-transformable. Thus, using

$$\phi(\alpha,\beta) = -(\alpha-\beta)^2 \tag{5.32}$$

in Corollary 5.19, one obtains

Theorem 5.25 The eigenvalues of a real matrix are all real (equivalently, A is aperiodic), if and only if

$$\text{Coef} \mid \eta I + A^2 \odot I + I \odot A^2 - 2A \odot A \mid \geq 0 .$$

Next, let us investigate $\phi(\alpha,\beta)$ given in (5.32). Clearly

$$\alpha,\beta \in \aleph \ ; \ \alpha \neq \beta \ \Rightarrow \text{Re}\,[\phi(\alpha,\beta)] < 0 .$$

On the other hand, if in the polynomial whose roots are $\phi(\lambda_i, \lambda_j) = -(\lambda_i-\lambda_j)^2$, all the coefficients are strictly positive, it follows that complex eigenvalues or repeated real eigenvalues are impossible. As a consequence we obtain

Theorem 5.26 Let $p(\lambda) = \sum_{i=0}^{n} p_i\lambda^i$, $p_n > 0$, be a real polynomial with roots $\{\lambda_i\}$. Let $q(\eta) = \sum_{i=0}^{k} q_i\eta^i$, $q_k > 0$, $k = \frac{1}{2}n(n-1)$, be a real polynomial with roots $-(\lambda_i - \lambda_j)^2$, $i = 2,3,\ldots,n$; $j = 1,2,\ldots,i-1$. For the roots of $p(\lambda)$ to be real and distinct, it is necessary and sufficient that $q_i > 0$, $i = 0,1,\ldots,k-1$.

It is interesting to note that this theorem is a limit case in our general theory. However, it was first presented by Waring in 1763, using a direct approach.

Combining Theorems 5.7 and 5.26 we have

Theorem 5.27 The eigenvalues of a real matrix A are all real and distrinct, if and only if

$$\text{Coef}\,|\,\eta I + A^2 \odot I + I \odot A^2 - 2A \odot A\,| > 0\,.$$

To close this section, we present a polynomial version for aperiodicity. This is a modification on a result obtained by Meerov in 1945.

Theorem 5.28 Consider the real polynomial $\Delta(\lambda) = \sum_{i=0}^{n} a_i \lambda^i$, $a_n > 0$. The roots of $\Delta(\lambda)$ are all real and distinct, if and only if Δ_{2n-1} below is positive innerwise (p.i.).

$$\Delta_{2n-1} = \begin{bmatrix} a_n & a_{n-1} \cdots & a_0 & 0 & \cdots & & 0 \\ & a_n & \cdots & a_1 & a_0 & \cdots & 0 \\ & & & & & & \vdots \\ & & a_n & a_{n-1} & a_{n-2} & & a_0 \\ & \bigcirc & 0 & na_n & (n-1)a_{n-1} & & a_1 \\ & & na_n & (n-1)a_{n-1} & (n-2)a_{n-2} & & 0 \\ & & & & & & \vdots \\ & na_n & & (n-1)a_{n-1} & \cdots & a_1 \cdots & 0 \\ na_n & (n-1)a_{n-1} & (n-2)a_{n-2} & \cdots & a_1 & \cdots & 0 \end{bmatrix} . \qquad (5.33)$$

NOTES AND REFERENCES

The key result of section 5.1 is Theorem 5.4 , due to Taub and Gutman [1] . It is based on the resultant of two polynomials. There are three ways to calculate the resultant: the first is based on the companion matrix and is traced back to MacDuffee [1]; the second is based on the Sylvester matrix and it dates back to 1840; the third is based on the Bezoutian and it dates back to 1764. For this and related topics consult Barnett [1]. The key result of Section 5.2 is Theorem 5.5, due to Stephanos [1]. See also MacDuffee [1] and Lancaster and Tismenetsky [1]. It is interesting to note that originally (1900), Stephanos posed the problem of composite polynomials, not composite matrices. He did so using (5.4) with $\phi_1 = 1$. Surprisingly, he did not use Sylvester matrix nor the Bezoutian, as we do. Rather, he constructed composite matrices and solved the polynomial version using the companion matrix. More surprisingly MacDuffee [1] quotes (1933), among many of Stephanos's results, the fact that "the resultant of $f(x) = 0$ and $g(x) = 0$ is $|A \otimes I - I \otimes B| = 0$, where f(x) is the characteristic equation of A and g(x) is the characteristic equation of B". However, years earlier (1907), the Sylvester matrix appeared in the book by Bocher [1]. It seems like neither Stephanos nor MacDuffee were aware of the Sylvester matrix or the Bezoutian. Theorems 5.8 - 5.12 are the consequence of Chapter 4 and Section 5.1, as mentioned in Taub and Gutman [1]. Likewise, Theorems 5.13 - 5.19 are the consequence of Chapter 4 and Section 5.2. Much of this discussion can be found in Gutman [1], [2], [4], Gutman and Chojnowski [1], and Gutman and Jury [1]. Theorem 5.28 is taken from Jury [1] and is due to Fuller [2]. Finally, we note that Fuller [1] was the first to use the Kronecker product in root clustering with respect to the left half plane. The notion of Transformability was explicitly defined only in Gutman [1].

Chapter 6 : SYMMETRIC MATRIX APPROACH

Recall the algebraic region $\aleph \subset \mathbb{C}$, defined by:

$$\aleph = \{(x+iy) : f(x,y) = \Sigma f_{ij}x^iy^j < 0\} . \tag{6.1}$$

This region has a complex representation

$$\aleph = \{\lambda \in \mathbb{C} : \phi(\lambda,\bar{\lambda}) < 0\} \tag{6.2}$$

where

$$\begin{aligned} &\phi(\alpha, \beta) = f(\frac{\alpha+\beta}{2}, \frac{\alpha-\beta}{2i}) = \sum \phi_{ij}\alpha^i\beta^j = L'(\alpha)\Phi L(\beta), \\ &L'(\alpha) = [1 \quad \alpha \quad \alpha^2 \dots], \\ &\Phi \triangleq [\phi_{ij}] \text{ is Hermitian.} \end{aligned} \tag{6.3}$$

Given $A \in \mathbb{C}^{n\times n}$ ***with spectrum*** $\sigma(A)$, ***and an algebraic region*** $\aleph$, ***we are looking for a criterion assuring*** $\sigma(A) \subset \aleph$.
For the left half plane, we are all familiar with the classical result of Lyapunov.

Theorem 2.8 $\sigma(A)$ lies in the left half plane, if and only if given any $Q = Q^*>0$ (p.d.), the (unique) solution $P = P^*$ of $PA + A^*P = -Q$ is p.d.

The unit circle version has the following form.

Theorem 2.9 $\sigma(A)$ lies in the unit circle, if and only if given any $Q = Q^* > 0$, the (unique) solution $P = P^*$ of $A^*PA-P = -Q$ is p.d.

A close look at the above matrix equations, reveals that they obey the rule

$$\Sigma\, \phi_{ij}A^iPA^{*j} = -Q . \tag{6.4}$$

Indeed, for the left half plane, we have

$$f(x,y) = x, \quad \phi(\alpha,\beta) = \alpha + \beta$$

so that

$$\phi_{10} = 1, \quad \phi_{01} = 1 .$$

Likewise, for the unit circle, we have

$$f(x,y) = -1 + x^2 + y^2 , \quad \phi(\alpha,\beta) = -1 + \alpha\beta$$

so that

$$\phi_{00} = -1, \quad \phi_{11} = 1 .$$

In the present chapter we search for the largest family of regions $\aleph$ for which the following theorem holds,

Theorem 6.0 $\sigma(A) \subset \aleph$, if and only if given any $Q = Q^* > 0$, the unique solution $P = P^*$ of $\Sigma\, \phi_{ij} A^i P A^{*j} = -Q$ is p.d.

6.1 MATRIX EQUATIONS — A SPECIAL CASE

In Section 3.8 we have used composite polynomials to discuss root clustering of a class of regions $\aleph$ satisfying Rank (Φ) - Signature (Φ) = 2. In Sections 4.3 and 4.8 we have shown that such a class is R-transformable. In Section 5.3, the criteria are based on composite polynomials and matrices. Here we investigate the same class using matrix equations.

Theorem 6.1 Suppose $\aleph$ satisfies Rank (Φ) - Sign (Φ) = 2, and consider $A \in C^{n\times n}$. Then $\sigma(A) \subset \aleph$, if and only if given any p.d. Hermitian matrix Q, the unique solution $P = P^*$ of

$$\Sigma\, \phi_{ij} A^i P A^{*j} = -Q \tag{6.4}$$

is p.d. .

Proof Sufficiency: Let z^* be the left eigenvector of A corresponding to the eigenvalue λ. Multiplying (6. 4) from left and right by z^* and z, respectively, and using $z^* A = \lambda z^*$, $A^* z = \bar{\lambda} z$, we obtain $\phi(\lambda, \bar{\lambda}) z^* P z = - z^* Q z$, where $\phi(\lambda, \bar{\lambda}) = \sum \phi_{ij} \lambda^i \bar{\lambda}^j = f(x, y)$.
Thus, if P and Q are p.d., it follows that $f(x,y) < 0$, and $\lambda \in \aleph$.

Necessity: This part consists of three main steps:

Step 1 Uniqueness of solution. In (6.4), the left hand side is linear in P. If we stack the rows of P and Q into column vectors p and q, respectively, (6. 4) is equivalent to $\phi(A \otimes \bar{A})p = -q$, where $\phi(A \otimes \bar{A}) = \sum \phi_{ij} A^i \otimes \bar{A}^j$. Thus, (6. 4) has a unique solution, if and only if $|\phi(A \otimes \bar{A})| = \prod_{i,j} \phi(\lambda_i, \bar{\lambda}_j) \neq 0$, or $\phi(\lambda_i, \bar{\lambda}_j) \neq 0\ \forall$ i, j. However, according to (4. 24) and (4. 28), $\lambda_i, \lambda_j \in \aleph \Rightarrow \phi(\lambda_i, \bar{\lambda}_j) \neq 0$. Thus, (6. 4) has a unique solution.

Step 2 There exists a p.d. pair $\{P_o, Q_o\}$ sativsfying (6.4). First note that using a similarity transformation on A,(6.4) is equivalent to $\Sigma\, \phi_{ij} A_1{}^i P_1 A_1{}^{*j} = -Q_1$, where $A_1 = T_1 A T_1{}^{-1}$, $P_1 = T_1 P T_1{}^*$, and $Q_1 = T_1 Q T_1{}^*$. Now, let J be the Jordan form of A. Then, $J = T_1 A T_1{}^1 = \Lambda + U$, where $\Lambda = \text{diag}[\lambda_1\ \lambda_2 \ldots\ \lambda_n]$ and the elements of U are 0 or 1, with all nonzero elements located on the diagonal above the main diagonal. For some small $\delta > 0$, define the nonsingular matrix T_2 by $T_2{}^{-1} = \text{diag}[1\ \ \delta\ \ \delta_2 \ldots \delta_{n-1}]$. Thus, $A_2 = T_2 T_1 A T_1{}^{-1} T_2{}^{-1} = \Lambda + \delta U$. Using $P_2 = -\text{diag}[\ldots \phi(\lambda_i, \bar{\lambda}_j) \ldots]$, we have $-Q_2 \triangleq \sum \phi_{ij} A_2^i P_2 A_2^{*j} = -\text{diag}[\ldots \phi^2(\lambda_i, \bar{\lambda}_i) \ldots] + O(\delta)$.
By hypothesis $\lambda_i \in \aleph\ \forall$ i. Thus P_2 is p.d.. By continuity, for small d, Q_2 is p.d. . Thus, there exist p.d. matrices P_o, Q_o, satisfying (6.4).

Step 3 Arbitrary p.d.Q. For an arbitrary p.d.Q , let $Q_t = tQ+(1-t)Q_o$, with $0 \le t \le 1$. Clearly Q_t is p.d. For $Q = Q_t$, (6.4) has a unique solution P_t which is continuous in t. Since P_o is p.d., it is left to prove that P_t never becomes singular. Recalling (4.24), we can write (6.4) in the form:

$$-\psi_1(A)P\psi_1(A)^* + \sum_{i=2} \psi_i(A)P\psi_i(A)^* = -Q. \tag{6.5}$$

Note that $\psi_1(A)$ is nonsingular. For if this is not true, there exists a nonzero vector x such that $x^*\psi_1(A) = 0$. Using P_o, Q_o in (6.5), and multiplying it from left and right by x^* and x, respectively, we obtain

$$\sum_{i=2} x^*\psi_1(A)P_o\psi_i(A)^*x = -x^*Q_o x.$$

The right hand side is negative, while the left hand side is nonnegative. This contradiction implies $|\psi_1(A)| \neq 0$. Now, we are in a position to prove $|P_t| \neq 0$. To this end, suppose that t_o is the first instant at which $P(t_o) = 0$. Thus, $P_{to} \ge 0$, and there exists $z \neq 0$ satisfying $z^*P_{to}z = 0$. Since $|\psi_1(A)| \neq 0$, there exists y satisfying $y^*\psi_1(A) = z^*$. Using P_{to}, Q_{to} in (6.5), and multiplying it from left and right by y^* and y, respectively, we obtain

$$\sum_{i=2} y^*\psi_i(A)P_{to}\psi_i(A)^*y = -y^*Q_{to}y.$$

The right hand side is negative, while the left hand side is nonnegative. This contradiction implies $P_t > 0$.

6.2 MOTIVATION FOR THE GENERAL CASE

Note first that the condition in Theorem 6.1 is ***sufficient*** for an ***arbitrary*** region. Indeed, let z_i^* be the left eigenvector of A corresponding to the eigenvalue λ_i. Multiplying (6.4) from left and right by z_i^* and z_i, respectively, and using $z_i^*A = \lambda_i z_i^*$, $A^*z_i = \bar{\lambda}_i z_i$, we obtain

$$\phi(\lambda_i, \bar{\lambda}_i)z_i^*Pz_i = -z_i^*Qz_i.$$

Thus, if Q and P are p. d., it follows that $\phi(\lambda_i, \bar{\lambda}_i) < 0$, and $\lambda_i \in \aleph\ \forall\ i$.

To discuss the necessity part we ***assume*** here that A is ***simple***; that is, it is diagonalizable via similarity transformation. In this case, simple calculations show that (6.4) is equivalent to

$$\sum \phi_{ij}\Lambda^i P_1 \Lambda^{*j} = -Q_1 \tag{6.6}$$

where

$$\Lambda = TAT^{-1} = \text{diag}[\lambda_1 \ \ \lambda_2 \ \ \dots \ \ \lambda_n]$$

$$P_1 = TPT^*, \qquad Q_1 = TQT^*.$$

The solution for P_1 is readily found to be

$$p_{ij} = -\frac{q_{ij}}{\phi(\lambda_i, \bar{\lambda}_j)} . \tag{6.7}$$

Since $P > 0 \Leftrightarrow P_1 > 0$, and $Q > 0 \Leftrightarrow Q_1 > 0$, we can study (6.6) - (6.7) instead of (6.4). Now that we have an explicit solution for P, we may find conditions under which $Q > 0$ implies $P > 0$. For this purpose, we recall an important result due to Schur.

Lemma 6.1 Let $A \in C^{n\times n}$ and $B \in C^{n\times n}$ be Hermitian. If $A > 0$ and $B \geq 0$ (p.s.d.) with all $b_{ii} > 0$, then

$$A \cdot B \stackrel{\Delta}{=} [a_{ij} b_{ij}] > 0 .$$

Proof Take $U = [u_{ij}]$ unitary such that $A = U^* \mathrm{diag}(\lambda_1, \ldots, \lambda_n) U$ where the λ_i are the eigenvalues of A. Thus we have $a_{ij} = \sum_{k=1}^{n} \overline{u_{kj}}\, u_{kj}$ and for $x = (x_1, \ldots, x_n)'$ we get

$$x^*(A \cdot B)x = \sum_{i,j=1}^{n} a_{ij}\, b_{ij}\, \overline{x_i}\, x_j$$

$$= \sum_{k=1}^{n} \lambda_k \Big(\sum_{i,j} b_{ij} (\bar{u}_{ki} \bar{x}_i)(u_{kj} x_j)\Big)$$

$$= \sum_{k=1}^{n} \lambda_k\, y_k^* B y_k$$

where $y_k = (u_{k1}x_1, \ldots, u_{kn}x_n)'$. Since $y_k^* B y_k \geq 0$ $(k = 1, \ldots, n)$ we get, setting

$$L \stackrel{\Delta}{=} \min_{1 \leq k \leq n} \lambda_k > 0 ,$$

$$x^*(A \cdot B)x \geq L \sum_{k=1}^{n} y_k^* B y_k = L \sum_{i,j} b_{ij}\, \overline{x_i}\, x_j \left(\sum_{k=1}^{n} \bar{u}_{ki}\, u_{kj}\right) =$$

$$= L(b_{11} | x_1|^2 + \ldots + b_{nn} | x_n|^2) .$$

Thus $x^*(A \cdot B)x > 0$ if $x \neq 0$.

Lemma 6.2 If B Hermitian is such that $A \cdot B \geq 0$ for all $A > 0$ then $B \geq 0$.

Proof Take $A = [a_{ij}]$ with $a_{ii} = \lambda$, $a_{ij} = 1$ if $1 \neq j$, and $\lambda \geq 1$. Then $A > 0$. Indeed for $x = (x_1, \ldots, x_n)' \neq 0$ we have

$$x^*Ax = \sum_{k=1}^{n} \overline{x}_k \, (x_1 + \ldots + x_n + (\lambda - 1)x_k)$$

$$= |x_1 + \ldots + x_n|^2 + (\lambda - 1) \sum_{k=1}^{n} |x_k|^2 > 0 .$$

Thus for all $\lambda > 1$ we have $A \cdot B \geq 0$, i.e.

$$\sum_{i \neq j} a_{ij} b_{ij} \, \overline{x_i} \, x_j \geq - \sum_{i=1}^{n} a_{ii} \, b_{ii} \, |x_i|^2 = - \lambda \sum_{i=1}^{n} b_{ii} |x_i|^2 ,$$

from where

$$x^*Bx = \sum_{i \neq j} a_{ij} b_{ij} x_i \, \overline{x_j} + \sum_{i=1}^{n} b_{ii} |x_i|^2 \geq (1 - \lambda) \sum_{i=1}^{n} b_{ii} |x_i|^2 .$$

Making $\lambda \to 1$ in this inequality we get $x^*Bx \geq 0$.

Now, suppose that $\aleph$ obeys the rule: $[-\phi^{-1}{}_{ij}] \in C^{n \times n}$ is p.s.d., and the $\lambda_i, \lambda_j \in \aleph$. Then, $-\phi(\lambda_i, \overline{\lambda}_i) > 0$. Thus, according to Lemma 6. 1 and (6. 7), $Q > 0$ implies $P > 0$. Conversely, according to Lemma 6.2 and (6.7), if for all $Q > 0$ we have $P > 0$, it follows that $[-\phi_{ij}^{-1}]$ is p. s. d.. We conclude by the following theorem.

Theorem 6.2 Let $A \in C^{n \times n}$ be simple with eigenvalues $\lambda_1, \ldots, \lambda_n$. Then $[-\phi^{-1}(\lambda_i, \lambda_j)]$ is positive semidefinite (p.s.d.), if and only if given any p.d. $Q = Q^*$, the unique solution $P = P^*$ of (6.4) is p.d..

Since we do not know $\lambda_1, \ldots, \lambda_n$, we are motivated to the following definition.

Definition 6.1 Region $\aleph$ is ***M-transformable***, if all $\alpha_i \in \aleph$ imply $F_\phi(\alpha_1, \ldots, \alpha_n) := [-\phi^{-1}(\alpha_i, \overline{\alpha}_j)]$ is p. s. d. .

Now, we can state a root clustering criterion for simple matrices.

Corollary 6.2 For $A \in C^{n \times n}$ simple and an M-transformable region $\aleph$, $\sigma(A) \subset \aleph$, if and only if given any p.d ., $Q = Q^*$, the unique solution $P = P^*$ of (6.4) is p.d. .

Does Theorem 6.2 hold for an arbitrary matrix A? How about Corollary 6.2? We answer these questions in Section 6.4. But first, we need some mathematical preparations.

6.3 SOME LEMMAS

Given $\phi(\alpha,\beta) = \Sigma \, \phi_{ij} \, \alpha^i \beta^j$ a Hermitian polynomial and $A \in C^{n \times n}$, we consider the linear operator

ϕ_A on C^{nxn} defined by

$$\phi_A(P) \triangleq \sum \phi_{ij} A^i P A^{*j} .$$

(6.8)

This operator is the main object of our study. Before getting to our main results we need a series of lemmas.

Lemma 6.3 Given $A \in C^{nxn}$ with eigenvalues $\lambda_1,...,\lambda_n$, the eigenvalues of ϕ_A, viewed as a linear operator on C^{nxn}, are the n^2 complex numbers $-\phi(\lambda_i, \bar{\lambda}_j)$, $i, j = 1, \ldots n$.

Proof If P_{k*} is the k-th row of $P \in C^{nxn}$ let P be the column vector of order n^2 given by $P\downarrow \triangleq (P_{1*} \ldots P_{n*})$. This is the **stacking operator**. Then, equation (6. 4), namely $\phi_A(P) = Q$, is equivalent to

$$-\phi(A, \bar{A})\, P\downarrow = Q\downarrow ,$$

(6.9)

where $\phi(A, \bar{A}) \triangleq \sum \phi_{ij} A^i \otimes \bar{A}^j$. This shows that $-\phi(A, \bar{A})$ is the matrix of ϕ_A in the basis $(M_{11},...,M_{ij},...,M_{nn})$ ordered in the lexicographic way, where the nxn matrix M_{ij} has only zero entries except the (i,j) entry which is 1. Therefore the characteristic polynomial of ϕ_A is $|\eta I + \phi(A, \bar{A})|$. Finally, according to Stephanos (Corollary 5. 5), the roots of this polynomial are the n^2 values $-\phi(\lambda_i, \bar{\lambda}_j)$.

Now let H_n be the set of nxn Hermitian matrices. Since $\phi_{ij} = \phi_{ji}$ we have $(\phi_A(P))^* = \phi_A(P^*)$. Thus $\phi_A(P)$ is Hermitian if $P = P^*$ and therefore ϕ_A is a linear operator on the n^2-dimensional real vector space H_n.

Lemma 6.4 If ϕ_A is nonsingular on H_n then it is nonsingular on C^{nxn}, i.e. $\phi(\lambda_i, \bar{\lambda}_j) \neq 0$ $(i, j = 1, \ldots, n)$.

Proof Assume that $\{P \in H_n: \phi_A(P) = 0\} = \{0\}$ and take $P \in C^{nxn}$, such that $\phi_A(P) = 0$. Then we have $\phi_A(P^*) = (\phi_A(P))^* = 0$, from where $\phi_A(P + P^*) = 0$ and also $\phi_A(iP\text{-}iP^*) = 0$. But $P + P^*$ and $i(P\text{-}P^*)$ are Hermitian, therefore $P + P^* = i(P - P^*) = 0$, i.e. $P = 0$. Thus ϕ_A is nonsingular on C^{nxn}, i.e. in view of Lemma 6.3, $\phi(\lambda_i,\lambda_j) \neq 0$ for all i,j.

Given E a vector space, let L(E) be the set of linear operators on E and Isom(E) the set of nonsingular linear operators (isomorphisms) on E.

Lemma 6.5 Given $\phi \in C_H[\alpha,\beta]$, the map $A \to \phi_A$ from C^{nxn} into $L(H_n)$ is continuous and if ϕ_A is nonsingular, then the map $B \to \phi_B^{-1}$ is defined and continuous in some neighborhood of A.

Proof For convenience we use the norms $\|M\| = \max_{x^*x=1} (x^*M^*Mx)^{1/2}$ for $M \in \mathbb{C}^{n\times n}$, and

$\|L\| = \max_{\|M\|=1} \|L(M)\|$ for $L \in L(\mathbb{C}^{n\times n})$. Thus we have $\|M\| = \|M^*\|$, $\|MN\| \le \|M\| \cdot \|N\|$ and

$\|L(M)\| \le \|L\| \cdot \|M\|$. Now given A, B and X in $\mathbb{C}^{n\times n}$, we have

$$\|\phi_A(X) - \phi_B(X)\| = -\sum \phi_{ij} A^i X A^{*j} + \sum \phi_{ij} B^i X B^{*j}$$

$$= -\sum \phi_{ij} (A^i - B^i) X A^{*j} - \sum \phi_{ij} B^i X (A^j - B^j)^*,$$

from where, since $\phi_{ij} = \bar{\phi}_{ij}$

$$\|\phi_A(X) - \phi_B(X)\| \le \|X\| \sum |\phi_{ij}| \, \|A^i - B^i\| \, (\|A\|^j + \|B\|^j).$$

We also have, using the symmetric polynomial S_p defined by (4.14),

$$\|A^i - B^i\| = \|\sum_{k=0}^{i-1} A^k (A-B) B^{i-1-k}\| \le \|A-B\| \, S_i(\|A\|, \|B\|).$$

Combining the last two inequalities, we get

$$\|\phi_A - \phi_B\| \le \|A-B\| \sum |\phi_{ij}| S_i\,(\|A\|, \|B\|)\,(\|A\|^j + \|B\|^j),$$

rom where $\phi_B \to \phi_A$ as $B \to A$, i.e. the map $A \to \phi_A$ is continuous.

When ϕ_A is nonsingular, since $\mathrm{Isom}(H_n)$ is an open subset of $L(H_n)$ there exists $\varepsilon > 0$ such that L is nonsingular if $\|L - \phi_A\| < \varepsilon$. On the other hand there exists $\eta > 0$ such that $\|\phi_B - \phi_A\| < \varepsilon$ if $\|A-B\| < \eta$. Thus $\phi_B \in \mathrm{Isom}(H_n)$ for all B in the open ball U: $\|B-A\| < \eta$, and the map $B \to \phi_B^{-1}$ is defined in U. This map is also continuous since it is composed of 2 continuous maps: $B \to \phi_B$ from U into $\mathrm{Isom}(H_n)$ and the inversion on $\mathrm{Isom}(H_n)$.

Lemma 6.6 Let E be a finite dimensional real vector space and L a linear operator on E such that L(E) has a nonempty interior. Then L is nonsingular.

Proof Let $\|\ \|$ be a norm on E. Since there exist $x_0 \in E$ and $r > 0$ such that $\{y \in E : \|y - L(x_0)\| \le r\} \subset L(E)$, we also have $\{y \in E : \|y\| \le r\} \subset L(E)$. Indeed $\|y\| \le r$ i.e. $\|y + L(x_0) - L(x_0)\| \le r$, implies that $y + L(x_0) = L(x)$ for some $x \in E$. Thus $y = L(x - x_0)$ and $y \in L(E)$. Now given any $y \ne 0$ we have $ry/\|y\| = L(x)$ for some $x \in E$, i.e. $y = L(x\|y\|/r)$. Thus $y \in L(E)$ and L is a surjective linear map, i.e., since the dimension of E is finite, an isomorphism.

Notation Let P_n be the set of positive definite nxn Hermitian matrices $P > 0$. For a given Hermitian polynomial ϕ, let $P_n(\phi)$ be the set of nxn complex matrices A such that $P_n \subset \phi_A(P_n)$, i.e.

$$(\forall Q > 0)\ (\exists P > 0)\ (\phi_A(P) = Q). \tag{6.10}$$

Lemma 6.7 Given $A \in \mathbb{C}^{n\times n}$ with eigenvalues $\lambda_1,\dots,\lambda_n$. If $A \in P_n(\phi)$ then ϕ_A is nonsingular, i. e. $\phi(\lambda_i, \bar{\lambda}_j) \ne 0$ $(i, j = 1, \dots n)$.

Proof Since P_n is an open subset of H_n our result is a direct application of Lemmas 6.6 and 6.4 .

Lemma 6.8 Given T a nonsingular nxn complex matrix, we have $A \in P_n(\phi)$ if and only if $TAT^{-1} \in P_n(\phi)$.

Proof Since $(T^*)^{-1} = (T^{-1})^*$ we have, setting $b \stackrel{\Delta}{=} TAT^{-1}$:

$$\phi_B(TPT^*) = -\sum \phi_{ij} B^i TPT^* B^{*j} = T\phi_A(P)T^* .$$

Assume that $A \in P_n(\phi)$ and take $Q > 0$. Since $T^{-1} Q(T^{-1})^* > 0$, there exists $P > 0$ such that $T^{-1} Q(T^{-1})^* = \phi_A(P)$, i.e. $Q = T\phi_A(P)T^* = \phi_B(TAT^*)$. But $TPT^* > 0$, hence $B \in P_n(\phi)$. The converse is now trivial .

Lemma 6.9 Given $A \in C^{n\times n}$ with eigenvalues $\lambda_1,\ldots,\lambda_n$ let $TAT^{-1} = \Lambda + U$ be a Jordan decomposition of A with T nonsingular, $\Lambda = \mathrm{diag}\ (\lambda_1,\ldots,\lambda_n)$ and $U = [u_{ij}]$ such that $u_{i,i+1} \in \{0,1\}$ and $u_{ij} = 0$ if $j \neq i+1$. Then given any scalar $\delta \neq 0$ we have $A \in P_n(\phi)$ if and only if $\Lambda + \delta U \in P_n(\phi)$.

Proof Consider the matrix $D \stackrel{\Delta}{=} \mathrm{diag}\,(1, \delta^{-1}, \ldots, \delta^{1-n})$. We have $DTA(DT)^{-1} = D(\Lambda+U)D^{-1} = \Lambda+DUD^{-1}$, and also $(DUD^{-1})_{ij} = \delta^{j-i}u_{ij} = \delta u_{ij}$ since $u_{ij} = 0$ for $j \neq i + 1$. Thus $DUD^{-1} = \delta U$, $DTA(DT)^{-1} = \Lambda+\delta U$ and we conclude by Lemma 6.8.

Lemma 6.10 Given $A \in C^{n\times n}$ with eigenvalues $\lambda_1,\ldots,\lambda_n$ and a Hermitian polynomial $\phi(\alpha,\beta)$, a necessary condition for $A \in P_n(\phi)$ is that $F_\phi\,(\lambda_1,\ldots,\lambda_n)$ is p.s.d.

In other words, if for all $Q > 0$ there exists $P > 0$ such that $\phi_A(P) = Q$, then $\phi(\lambda_i, \bar{\lambda}_j) \neq 0$ $(i, j = 1,\ldots, n)$ and $[\phi^{-1}(\lambda_i, \bar{\lambda}_j)]$ is p. s. d..

Proof Let $TAT^{-1} = \Lambda + U$ be the Jordan decomposition of A. By Lemma 6.9 we have for all $\delta \neq 0$, $\Lambda + \delta U \in P_n(\phi)$. Thus, for all $Q > 0$ and $\delta \neq 0$ there exists $P_\delta > 0$ such that $Q = \phi_{\Lambda+\delta U}(P_\delta)$. But ϕ_Λ is nonsingular and $\phi(\lambda_i, \bar{\lambda}_j) \neq 0$ (Lemma 6. 7). Thus, ϕ_Λ and $\phi_{\Lambda+\delta U}$ are also nonsingular (Lemma 6. 3). Hence $P_\delta = \phi^{-1}_{\Lambda+\delta U}(Q)$ and by Lemma 6. 5 P_δ has a limit as $\delta \to 0$, namely $\lim P_\delta = \phi^{-1}_\Lambda(Q) = P$. Thus, $Q = \phi_\Lambda(P) = -\sum \phi_{ij} \Lambda^i P \bar{\Lambda}^j$ and we are back at (10). We conclude by Theorem 6. 2.

In order to get a converse to Lemma 6.10, we need the following Lemma.

Lemma 6.11 Given a Hermitian polynomial $\phi(\alpha,\beta)$ and $A \in C^{n\times n}$, a necessary (and sufficient) condition for $\sigma(A) \subset \aleph$ is that there exists $P > 0$ such that $\phi_A(P) > 0$.

Note that this result is not a root clustering criterion since we do know how to select P.

Proof Identical to Step 2 in the proof to Theorem 6.1.

6.4 ROOT CLUSTERING CRITERIA

In this section we combine our results to form a root clustering criterion.

Theorem 6.3 Given a Hermitian polynomial $\phi(\alpha,\beta)$ and $A \in C^{n\times n}$ with eigenvalues $\lambda_1,\ldots,\lambda_n$, any of the following conditions assures that $A \in P_n(\phi)$ (see (6.10))

(a) $F_\phi(\lambda_1,\ldots,\lambda_n) \geq 0$ and A is simple.

(b) $F_\phi \geq 0$ in some neighborhood of $(\lambda_1,\ldots,\lambda_n)$.

(c) $F_\phi(\lambda_1,\ldots,\lambda_n) > 0$.

(d) $F_\phi(\lambda_1,\ldots,\lambda_n) \geq 0$ and ϕ is M-transformable.

Conversely, $A \in P_n(\phi)$ implies that $F_\phi(\lambda_1,\ldots,\lambda_n) \geq 0$.

However, ***we do not know whether*** $F_\phi(\lambda_1,\ldots,\lambda_n) \geq 0$ ***alone implies*** $A \in P_n(\phi)$. In other words, Theorem 6.2 was proved for simple matrices, can be extended to arbitrary matrices only in the necessity part (Lemma 6.10) but not in the sufficiency part.

Proof (a) This is part of Theorem 6.2. (b) First we show that $(\forall Q > 0)(\exists P \geq 0)(\phi_A(P)= Q)$. For that let $\{A_k\}$ be a sequence of simple matrices converting to A. At the price of a reordering we can assume that the eigenvalues $\lambda_i^{(k)}$ of A_k converge to λ_i $(i = 1,\ldots,n)$. We set $\lambda^{(k)} \triangleq (\lambda_1^{(k)}, \ldots, \lambda_n^{(k)})$ and $\lambda \triangleq (\lambda_1, \ldots, \lambda_n)$. Thus for k large enough we have $F_\phi(\lambda^{(k)}) \geq 0$, and according to (a), given any $Q > 0$, there exists $P_k > 0$ such that $Q = \phi_{A_k}(P_k)$, i. e., since ϕ_{A_k} is nonsingular (Lemma 6.7), $P_k = \phi_{A_k}^{-1}(Q)$. Since ϕ_A is also nonsingular $(\phi(\lambda_i, \lambda_j) \neq 0)$ we can define $P \triangleq \phi_A^{-1}(Q)$. By Lemma 6. 5 we have $\phi_{A_k}^{-1} \to \phi_A^{-1}$, and also $\phi_{A_k}^{-1}(Q)$, i. e. $P_k \to \phi_A^{-1}(Q)$, i. e. $P_k \to P$ as $k \to \infty$. Now we show that in fact $P = \phi_A^{-1}(Q)$ is positive definite. Since $F_\phi(\lambda) \geq 0$, we have $\lambda_1,\ldots,\lambda_n \in \aleph$ (the diagonal elements of $F_\phi(\lambda)$ are positive !) and thus, by Lemma 6.11, there exists $P_0>0$ such that $Q_0 \triangleq \phi_A(P_0) > 0$. If $P_0 = P$, we are done. If not, then $Q \neq Q_0$ and we consider for $t \in [0, 1]$, the matrices $Q_t \triangleq (1-t)Q_0 + tQ$ and $P_t \triangleq \phi_A^{-1}(Q_t) = (1-t)P_0 + tP$. Thus for all $t \in [0, 1]$, since $P \geq 0$, we have $P_t > 0$.
Hence for all $Q_2 > 0$ we have $\phi_A^{-1}([Q_0, Q_2)) \subset P_n$. Since P_n is an open subset of H_n, the $n\times n$ Hermitian matrices, there exists $\rho > 0$ such that $M > 0$ as soon as $\| Q - M \| < \rho$. Thus $Q_2 \triangleq Q/t - (1/t - 1)Q_0$ is p. d. if we choose

$$\frac{\| Q - Q_0 \|}{\rho + \| Q - Q_0 \|} < t < 1.$$

Since now $Q \in [Q_0, Q_2)$, we have $P = \phi_A^{-1}(Q) > 0$.
(c) $\Rightarrow$ (b) is obvious. To prove (d) $\Rightarrow$ (b), note that if $F_\phi(\lambda) \geq 0$, then $\lambda_1,\ldots,\lambda_n \in \aleph$ and since $\aleph$ is open and M-transformable, we have $F_\phi \geq 0$ in some neighborhood of $(\lambda_1,\ldots,\lambda_n)$. Finally, the converse statement is Lemma 6.10.

Theorem 6.4 Consider an M-transformable region $\aleph$ and $A \in C^{n\times n}$. We have $\sigma(A) \subset \aleph$ if and only if for all $Q > 0$ there exists $P > 0$ satisfying $\Sigma\, \phi_{ij} A^i P A^{*j} = -Q$.

Proof Sufficiency holds for an arbitrary region as stated at the beginning of Section 6.2. Necessity is a direct application of Definition 6.1 and Theorem 6.3 (d).

Note that in the theorem we state ***for all*** $Q > 0$. However, if we go back to the proof of the sufficiency part, we see that it is sufficient that ***given any*** $Q > 0$ there exists $P > 0$. For the necessity part, it is clear from Theorem 6.3 that if (d) implies $A \in P_n(\phi)$, it implies, as a special case, that given any $Q > 0$ there exists $P > 0$. Thus, we have the following root clustering criterion.

Theorem 6.5 Consider an M-transformable region $\aleph$ and $A \in C^{n\times n}$. We have $\sigma(A) \subset \aleph$ if and only if given any $Q > 0$ there exists $P > 0$ satisfying $\sum \phi_{ij} A^i P A^{*j} = -Q$

Previously we have presented root clustering criteria based on ***polynomial inequalities*** in the entries of A. On the other hand, Theorem 6.5 generates ***rational inequalities*** in the entries of A. This is so, since from (6.9)

$$P = (-\phi^{-1}(A, \bar{A})Q\downarrow)\uparrow \tag{6.11}$$

where $\uparrow$ is the ***inverse stacking*** operator. In other words

$$P = |-\phi(A, \bar{A})|^{-1}(\mathrm{adj}(-\phi(A, \bar{A}))Q\downarrow)\uparrow . \tag{6.12}$$

We are now ready to state a polynomial version of Theorem 6.5.

Theorem 6.6 Consider an M-transformable region $\aleph$ and $A \in C^{n\times n}$. We have $\sigma(A) \subset \aleph$ if and only if given any $Q > 0$, the following n+1 polynomial inequalities (in the entries of A) are satisfied.

(i) $(-1)^n |\phi(A, \bar{A})| > 0$

(ii) $(-1)^{n+1}(\mathrm{adj}(\phi(A, \bar{A}))Q\downarrow)\uparrow$ is p. d. .

We are aware of the possibility to solve (6.9) using rows operations. Yet, in case A is a function of parameters, we wish to present explicit inequalities.

Proof Let $\lambda_1,\dots,\lambda_n$ be the eigenvalues of A. As mentioned earlier, the eigenvalues of $\phi(A, \bar{A})$ are the n^2 values $\phi(\lambda_i, \bar{\lambda}_j)$. Thus, since $\phi(\alpha, \beta)$ is Hermitian,

$$|\phi(A, \bar{A})| = \prod_{1\le i,j\le n} \phi(\lambda_i, \bar{\lambda}_j) = \prod_{1\le i\le n} \phi(\lambda_i, \bar{\lambda}_i) \prod_{1\le i,j\le n} |\phi(\lambda_i, \bar{\lambda}_j)|^2. \tag{6.13}$$

Thus, (i) is necessary for $\sigma(A) \subset \aleph$ when $\phi(\alpha,\beta) \neq 0$ for all $\alpha,\beta \in \aleph$, in particular, when $\aleph$ is M–transformable. Now, we can multiply (6. 12) by $|-\phi(A, \bar{A})| > 0$ and apply directly Theorem 6.5. This completes the proof.

At the opening of the chapter we were looking for the largest family of regions for which Theorem 6.0 holds.

Theorem 6.7 The largest family of regions $\aleph$ for which Theorem 6.0 holds is the family of M-transformable regions.

Proof Suppose $\lambda_1,\ldots,\lambda_n \in \aleph$ and let $A = \mathrm{diag}(\lambda_1,\ldots,\lambda_n)$. Then, $A \in P_n(\phi)$ and by Lemma 6.10 $F_\phi(\lambda_1,\ldots,\lambda_n) \geq 0$. Since $\lambda_i \in \aleph$ are arbitrary, $\aleph$ is M-transformable.

6.5 M-TRANSFORMABILITY

As mentioned in Section 6.4, Theorem 6.0 applies to M-transformable regions; therefore, it is important to be able to test it. However, from Definiton 6.1, M-transformability depends on n, the order of matrix A, while we expect it to be a region's property. In the present section we show that a region satisfying Rank (Φ) - Signature (Φ) = 2 is M-transformable. In fact we generate a more general family contained in M-transformability. First recall (Section 4.3) that if $\aleph$ has the property Rank (Φ) - Signature (Φ) = 2, it has the form

$$\text{(i)}\quad \phi(\alpha,\beta) = -\phi_1(\alpha)\overline{\phi_1(\beta)} + \sum_{i=2} \phi_i(\alpha)\overline{\phi_i(\beta)}$$

$$\text{(ii)}\quad \aleph = \left\{ \lambda \in C\colon \phi(\lambda,\bar{\lambda}) = -|\phi_1(\lambda)|^2 + \sum_{i=2} |\phi_i(\lambda)|^2 < 0 \right\}. \tag{6.14}$$

Now, consider the more general family

$$\text{(i)}\quad \phi(\alpha,\beta) = -\left(\phi_1(\alpha)\overline{\phi_1(\beta)} - \sum_{i=2} \phi_i(\alpha)\overline{\phi_i(\beta)}\right)^{2m+1} + \sum_{j=1} \psi_j(\alpha)\overline{\psi_j(\beta)}$$

$$\text{ii)}\quad \aleph = \left\{ \lambda \in C\colon \phi(\lambda,\bar{\lambda}) = -\left(|\phi_1(\lambda)|^2 - \sum_{i=2} |\phi_i(\lambda)|^2\right)^{2m+1} + \sum_{j=1} |\psi_j(\lambda)|^2 < 0 \right\}. \tag{6.15}$$

Note that in case $m = 0$, (6.15) reduces to (6.14). We claim that $\aleph$ given by (6.15) is M-transformable. To prove this we need some preparation.

Lemma 6.12 Matrix $A = [a_i \ \bar{a}_j]$ is p. s. d.

Proof Let $a = \mathrm{col}[a_1 \ldots a_n]$. Then $A = aa^*$. Now, given any $v \in C^n$, $v^*Av = v^*aa^*v = |v^*a|^2 \geq 0$.

Lemma 6.13 Let $M = [m_{ij}]$ be Hermitian p. s. d. such that $m_{ii} < 1 \forall i$. Then the matrix $\left[\frac{1}{1-m_{ij}}\right]$ is p. s. d.

Proof First note that if $M = M^*$ and $|m_{ii}| < 1$, it follows that $|m_{ij}| < 1$. This is so, since every 2×2 principle minor is nonnegative; or, $m_{ii}m_{jj} - |m_{ij}|^2 \geq 0$, and $|m_{ij}|^2 \leq m_{ii}m_{jj} < 1$.
Thus, we can write $\frac{1}{1-m_{ij}} = \sum_{k=0}^{\infty} m_{ij}^k$, or in matrix form $\left[\frac{1}{1-m_{ij}}\right] = \sum_{k=0}^{\infty} [m_{ij}^k]$. Since $[m_{ij}]$ is p.s.d. it follows that, the Schur (term by term) product, $[m_{ij}^k]$, is also p.s.d. Now, given any $v \in \mathbf{C}^n$, $v^*[m_{ij}^k]v \geq 0$, and we have $v^*\left[\frac{1}{1-m_{ij}}\right]v = \sum_{k=0}^{\infty} v^*[m_{ij}^k]v \geq 0$, as desired.

From (6. 15) we have $-\left(|\phi_1(\lambda)|^2 - \sum_{i=2} |\phi_i(\lambda)|^2\right)^{2m+1} < 0$, or $|\phi_1(\lambda)|^2 - \sum_{i=2} |\phi_i(\lambda)|^2 > 0$.

Thus,

$$\begin{aligned} &\text{(i)} \quad -|\phi_2(\lambda)|^2 + \sum_{i=2} |\phi_i(\lambda)|^2 < 0 \qquad \forall\ \lambda \in \aleph \\ &\text{p. s. d.,} \\ &\text{(ii)} \quad \sum_{i=2} \left|\frac{\phi_i(\lambda)}{\phi_1(\lambda)}\right|^2 < 1 \qquad \forall\ \lambda \in \aleph . \end{aligned} \tag{6.16}$$

Remark 6.1 Relation (6.16) implies that $\aleph$ given by (6.15) is bounded by a specific $r(\Phi) - s(\Phi) = 2$ region.

Lemma 6.14 $\phi_1(\lambda_i)\overline{\phi_1(\lambda_j)} - \sum_{k=2} \phi_k(\lambda_i)\overline{\phi_k(\lambda_j)} \neq 0 \quad \forall\ \lambda_i \in \aleph, \quad i = 1, 2, \ldots . n.$

Proof Let

$$r_{ij} = \sum_k \frac{\phi_k(\lambda_i)\overline{\phi_k(\lambda_j)}}{\phi_1(\lambda_i)\overline{\phi_1(\lambda_j)}} .$$

By Lemma 6.12, matrix $R = [r_{ij}]$ is p.s.d. From (6.16), this matrix has a principle diagonal satisfying $r_{ij} < 1$. Thus, as in the proof of Lemma 6.13, we have $|r_{ij}| < 1$, or $1 - r_{ij} \neq 0$. Multiplying $(1 - r_{ij})$ by $\phi_1(\lambda_i)\overline{\phi_1(\lambda_j)}$, our claim follows.

Following Lemma 6. 14, we can define

$$m_{ij} = \frac{\sum_{t=1} \psi_t(\lambda_i)\overline{\psi_t(\lambda_j)}}{\left(\phi_1(\lambda_i)\overline{\phi_1(\lambda_j)} - \sum_{k=2} \phi_k(\lambda_i)\overline{\phi_k(\lambda_j)}\right)^{2m+1}} \tag{6.17}$$

$$\tilde{\phi}(\lambda_i, \bar{\lambda}_j) = -1 + m_{ij} \;. \tag{6.18}$$

Lemma 6.15 The matrix $[m_{ij}]$ is p.s.d. in $\aleph$.

Proof In the proof of Lemma 6.14, we have seen that in $\aleph$, R is p.s.d, with $|r_{ii}| < 1 \;\forall$ i. Thus, according to Lemma 6.13, $\left[\frac{1}{1-r_{ij}}\right]$ is p. s. d. in $\aleph$. Simple algebra shows

$$[m_{ij}] = \sum_{t=1} \left[\psi_t(\lambda_i)\overline{\psi_t(\lambda_j)}\right] \cdot \left(\left[\frac{1}{\phi_1(\lambda_i)\phi_1(\lambda_j)}\right] \cdot \left[\frac{1}{1-r_{ij}}\right]\right)^{[2m+1]} \geq 0,$$

where $\cdot$ stands for Schur (term by term) product, and $A^{[2]} = A \cdot A$.

Next, from (6.15), $m_{ii} < 1$ in $\aleph$. Since $[m_{ij}]$ is p.s.d., it follows that $|m_{ij}| < 1$ in $\aleph$. Thus we may define in $\aleph$

$$\tilde{\psi} = \left[-\frac{1}{\tilde{\phi}(\lambda_i, \bar{\lambda}_j)}\right]; \quad \psi = \left[-\frac{1}{\phi(\lambda_i, \bar{\lambda}_j)}\right]. \tag{6.19}$$

Theorem 6.8 Region $\aleph$, given by (6.15), is M-transformable.

Proof According to our previous results, $\tilde{\psi} = \left[\frac{1}{1-m_{ij}}\right] \geq 0$ in $\aleph$. Since

$$\psi = \left(\left[\frac{1}{\phi_1(\lambda_i)\phi_1(\lambda_j)}\right] \cdot \left[\frac{1}{1-r_{ij}}\right]\right)^{[2m+1]} \cdot \tilde{\psi},$$

it follows that ψ is p.s.d. in $\aleph$, as desired.

Remark 6.2 In light of (6.18) and the fact $|m_{ij}| < 1$, it follows that in $\aleph$, $\mathrm{Re}[\tilde{\phi}] < 0$. Thus, $\aleph$ given by (6.15) is R-transformable.

Example 6.1

In (6), let $\phi_1(\lambda) = 1$, $\phi_2(\lambda) = \lambda$, $\psi_1(\lambda) = \lambda + \frac{1}{2}$. Thus

$$\phi(\alpha, \beta) = -(1-\alpha\beta)^3 + (\alpha + \tfrac{1}{2})(\beta + \tfrac{1}{2})$$

$$\aleph = \left\{ x + iy : -(1-x^2-y^2)^3 + (x+\tfrac{1}{2})^2 + y^2 < 0 \right\}.$$

Note that

$$\Phi = \begin{bmatrix} -\frac{3}{4} & \frac{1}{2} & 0 & 0 \\ \frac{1}{2} & 4 & 0 & 0 \\ 0 & 0 & -3 & 0 \\ 0 & 0 & 0 & 1 \end{bmatrix}$$

and that

$$r(\Phi) - s(\Phi) = 4.$$

However, by construction, $\aleph$ is M-transformable. Note that by (6.16) the above region is bounded by the unit disk.

6.6 POLYNOMIAL ROOT CLUSTERING

In previous sections we have discussed root clustering (spectrum inclusion) of square matrices. In many applications we are given a polynomial rather than a matrix. As in previous chapters, we wish now to develop a polynomial version to the Linear Matrix Equations. One may argue that in Theorem 6.5 we can use a companion matrix of the given polynomial. However, to simplify calculations, we prefer polynomial rather than matrix operations. Moreover, we wish to discard matrix equations, if possible. Let

$$\Delta(\lambda) = \sum_{i=0}^{n} a_i \lambda^i \quad ; \quad a_n = 1 \tag{6.20}$$

be a polynomial with $\lambda_1, \lambda_2, \ldots, \lambda_n$ as roots. Let a region $\aleph$ in the complex plane be defined by (6.1) - (6.3). We know (Section 5.1) that the polynomial

$$q(\eta) \triangleq \sum_{m=0}^{n^2} q_m \eta^m = \operatorname{Res}[\,\Delta(\lambda),\ \operatorname{Res}(\bar{\Delta}(s), \eta - \phi(\lambda, s))] \tag{6.21}$$

has $\phi(\lambda_i, \bar{\lambda}_j)$ as roots. By construction,

$$\sum_{m=0}^{n^2} q_m \phi^m(\lambda_i, \bar{\lambda}_j) = 0 \quad ; \quad q_{n^2} \neq 0$$

$$q_0 \neq 0 \quad \Leftrightarrow \quad \phi(\lambda_i, \bar{\lambda}_j) \neq 0 \ \forall \ i \tag{6.22}$$

Solving (2.22) for q_0,

$$q_0 = -\sum_{m=1}^{n^2} q_m \phi^m(\lambda_i, \bar{\lambda}_j) = -\phi(\lambda_i, \bar{\lambda}_j) \sum_{m=1}^{n^2} q_m \phi^{m-1}(\lambda_i, \bar{\lambda}_j).$$

Multiplying by $(TT^*)_{ij} = \sum_{k=1}^{n} \lambda_i^{k-1} \bar{\lambda}_j^{k-1}$, where T is the Vandermonde matrix, we have

$$q_0(TT^*)_{ij} = -\phi(\lambda_i, \bar{\lambda}_j)\left(\sum_{m=1}^{n^2} q_m \phi^{m-1}(\lambda_i, \bar{\lambda}_j)\right)\left(\sum_{k=1}^{n} \lambda_i^{k-1}\bar{\lambda}_j^{k-1}\right). \tag{6.23}$$

Define $C = [c_{kl}]$ by

$$\left(\sum_{m=1}^{n^2} q_m \phi^{m-1}(\lambda, s)\right)\left(\sum_{k=1}^{n} (\lambda s)^{k-1}\right) \overset{\text{mod}}{=} \sum_{k,m=1}^{n} c_{km} \lambda^{k-1} s^{m-1} \tag{6.24}$$

where $\overset{\text{mod}}{=}$ means "λ, s modulo $\Delta(\cdot)$, $\bar{\Delta}(\cdot)$", respectively.

Substituting (6.24) in (6.23), recalling that $\{\lambda_i\}$ are the roots of $\Delta(\lambda)$, we find

$$q_0(TT^*)_{ij} = -\phi(\lambda_i, \bar{\lambda}_j) \sum_{k,m=1}^{n} c_{km} \lambda_i^{k-1} \bar{\lambda}_j^{m-1}$$

and in matrix form

$$q_0(TT^*) = [-\phi(\lambda_i, \bar{\lambda}_j)] \cdot TCT^* \tag{6.25}$$

where $\cdot$ is the Schur (term by term) product.

Remark 6.3 Practically, we take the left hand side of (6.24) modulo $\Delta(\cdot)$ and generate directly $[c_{km}]$.

Remark 6.4 In case $\Delta(\lambda)$ is a ***real polynomial,*** and $\aleph$ ***symmetric*** we can use arguments similar to Corollary 5.8. Then we replace $q(\eta)$ in (6.21), by

$$\tilde{q}(\eta) = \sum_{m=0}^{\frac{1}{2}n(n+1)} \tilde{q}_m \eta^m, \text{ given by}$$

$$\tilde{q}(\eta) = \hat{q}(\eta)(q(\eta)/\hat{q}(\eta))^{\frac{1}{2}} \tag{6.26}$$

where $q(\eta)$ is given in (6.21), and

$$\hat{q}(\eta) = \text{Res}[\Delta(\lambda), \eta - \phi(\lambda, \lambda)] . \tag{6.27}$$

Now, we are ready to state our main result.

Theorem 6.9 Let $\Delta(\lambda)=\sum_{i=0}^{n} a_i\lambda^i$ and consider an M−transformable region $\aleph$. Let C be constructed via (6.21), (6.24), and in the real symmetric case via (6.24), (6.26), (6.27). Then $\sigma(\Delta)\subset\aleph$, if and only if $q_o C$ is p.d.

To prove the theorem, we need the following result.

Lemma 6.16 Let A and $B=[b_{ij}]$ be p.s.d. Hermitian matrices. Then

$$|A\cdot B|\geq|A|\prod_i b_{ii}.$$

Proof of Theorem 6.9 Necessity. If $\lambda_1,\dots,\lambda_n\in\aleph$, an M−transformable region, then $\phi(\lambda_i,\bar\lambda_j)\neq 0\ \forall\, i, j$; thus $q_o\neq 0$. Also, $\left[-\phi^{-1}(\lambda_i,\bar\lambda_j)\right]$ p. s. d. implies $\phi(\lambda_i,\bar\lambda_i)<0\ \ \forall\, i$. Thus, from (6. 25)

$$T(q_oC)T^*=TT^*\cdot\left[\frac{q_o^2}{-\phi(\lambda_i,\bar\lambda_j)}\right]\qquad\text{is p. s. d.},$$

and by Lemma 6.16

$$|T^2|\ |q_0C|=\left|TT^*\cdot\left[\frac{q_0^2}{-\phi(\lambda_i,\bar\lambda_j)}\right]\right|\geq|T^2|\prod_i\frac{q_o^2}{-\phi(\lambda_i,\bar\lambda_i)}$$

so, for distinct $\lambda_1,\dots,\lambda_n\in\aleph$, for which $|T|\neq 0$

(i) $\quad |q_oC|\geq\prod_i\dfrac{q_o^2}{-\phi(\lambda_i,\bar\lambda_i)}>0,$

(ii) $\quad q_oC$ is p.s.d. .

From the continuity of q_oC with respect to $\lambda_1,\dots,\lambda_n$, (i) and (ii) hold for all $\lambda_1,\dots,\lambda_n\in\aleph$. Thus q_0C is nonsingular and p.s.d.; thus p.d. in $\aleph$.

Sufficiency. If q_0C is p. d., then $q_0\neq 0$ and $\phi(\lambda_i,\bar\lambda_j)\neq 0\ \forall\, i, j$. Thus from (6. 25)

$$(TT^*)\cdot\left[\frac{q_0^2}{-\phi(\lambda_i,\bar\lambda_j)}\right]=Tq_0CT^*\qquad\text{is p. s. d.}$$

and the diagonal elements are nonnegative,

$$(TT^*)_{ii}\cdot\frac{q_0^2}{-\phi(\lambda_i,\bar\lambda_i)}\geq 0\ .$$

But $q_0^2 > 0$, and $(TT^*)_{ii} = \sum_{m=0}^{n-1} |\lambda_i|^{2m} > 1 > 0$; thus, $\phi(\lambda_i, \bar{\lambda}_i) < 0$, as desired.

Since q_0 in (6.21) satisfies

$$q_0 = \prod_{i,j=1}^{n} (-\phi(\lambda_i, \bar{\lambda}_j)) = \prod_{i=1}^{n} -\phi(\lambda_i \bar{\lambda}_i) \prod_{i<j} |\phi(\lambda_i, \bar{\lambda}_j)|^2, \tag{6.28}$$

it follows that

$$q_0 > 0 \quad \text{provided} \quad \sigma(\Delta) \subset \aleph . \tag{6.29}$$

Thus, we can simplify Theorem 6.9 as follows.

Corollary 6.9 Let $\Delta(\lambda) = \sum_{i=0}^{n} a_i \lambda^i$ and consider an M–transformable region $\aleph$. Let C be constructed via (6.21), (6.24). Then $\sigma(\Delta) \subset \aleph$, if and only if (i) $q_0 > 0$, and (ii) C is p.d.

In case n is even, $q_0 C$ can be replaced by C.

Theorem 6.10 Let $\Delta(\lambda) = \sum_{i=0}^{n} a_i \lambda^i$, with n even, and consider an M–transformable region $\aleph$. Let C be constructed via (6.21), (6.24). Then $\sigma(\Delta) \subset \aleph$, if and only if C is p.d.

Proof Necessity is evident by Corollary 6.9, part (ii). It holds for an arbitrary n.
Sufficiency: for each i,j, we can write

$$q_0 = \prod_{k,m} -\phi(\lambda_k, \bar{\lambda}_m) = -\phi(\lambda_i, \bar{\lambda}_j) \prod_{\substack{(k,m)\neq \\ (i,j)}} -\phi(\lambda_k, \bar{\lambda}_m)$$

Set

$$Q_{ij} = \prod_{\substack{(k,m)\neq \\ (i,j)}} -\phi(\lambda_k, \bar{\lambda}_m), \quad \text{and} \quad \phi_{ij} = \phi(\lambda_i, \bar{\lambda}_j).$$

Then, $q_0 = -\phi_{ij} Q_{ij}$. Substituting in (6.25), we obtain

$$[-\phi_{ij} Q_{ij}] \cdot TT^* - [-\phi_{ij}] \cdot TCT^* = 0$$

and using $Q = [Q_{ij}]$, $\Phi = [\phi_{ij}]$, we have

$$\Phi \cdot Q \cdot TT^* - \Phi \cdot TCT^* = 0$$

or

$$\Phi \cdot [Q \cdot TT^* - TCT^*] = 0 .$$

By a continuity argument it can be proved (details are omitted) that in fact

$$Q \cdot TT^* = TCT^* \tag{6.30}$$

For distinct $\{\lambda_i\}$, T is nonsingular, so that for C p. d, TCT^* is p. d., and from (6. 30), $Q \cdot TT^*$ is p. d. In particular, the main diagonal is positive, $Q_{ii} \sum_{k=0}^{n-1} \lambda_i^{2k} > 0$, or $Q_{ii} > 0$.

For $i = 1$,

$$Q_{11} = \prod_{\substack{(k, m) \neq \\ (1, 1)}} -\phi_{km} = \prod_{k<m} |-\phi_{km}|^2 \prod_{i=2}^{n} (-\phi_{ii}) > 0 .$$

Thus

$$\phi_{ij} \neq 0 \qquad \forall (i, j) \neq (1, 1) .$$

Likewise, for $i = 2$,

$$\phi_{ij} \neq 0 \qquad \forall (i, j) \neq (2, 2) .$$

We conclude that $\phi_{ij} \neq 0 \;\; \forall \;\; i, j$, and $q_0 \neq 0$. Now, we can divide (6. 25) by ϕ_{ij} to obtain

$$\left[\frac{q_0}{-\phi_{ij}}\right] \cdot TT^* = TCT^* > 0$$

and the main diagonal

$$\frac{q_0}{-\phi_{ii}} [TT^*]_{ii} > 0 .$$

Since $[TT^*]_{ii} > 0$, it follows that

$$\frac{q_0}{-\phi_{ii}} > 0 \tag{6.31}$$

From (6.28) we see that for n even, $q_0 > 0$; thus,

$$\phi_{ii} < 0 \quad \text{and} \quad \lambda_i \in \aleph \quad \forall i .$$

For arbitrary λ_i, we construct a series of simple polynomials converging to $\Delta(\lambda)$. It can be shown that $\phi_{ii} < 0$ still holds.

Discussion

Now, we are ready to comment on the structure of our new root clustering criterion.First note that we make use of the indeces $\{q_i\}$. Although these indeces play an important rolein P-Transformability (Theorem 5.8), our approach cannot be considered as a reduction of the inequalities in Theorem 5.8. This is so despite the fact that in the intersection of P and M-Transformability one might draw such a conclusion. We feel that our approach is the polynomial version of Theorem 6.5.Indeed, operating modulo Δ on (6.25), we obtain a set of n^2 linear equations in the unknown c_{ij}'s. However, instead of solving this set, we use modular arithmetic, according to (6.24), and get the c_{ij}'s directly. The use of $\{q_i\}$ in our approach should not surprise the reader since, these are the characteristic coefficients of matrix $\phi(A \otimes \bar{A})$ where A is the companion matrix of $\Delta(\cdot)$. How are our results related to known criteria, like Hermitian form ? To answer this question we present a simple example. Consider a second order polynomial and the left half plane,

$$\Delta(\lambda) = \lambda^2 + a\lambda + b$$
$$\aleph = \{x + iy:\ x < 0\}\ .$$

Hermite criterion has the form

$$H = \begin{bmatrix} ab & 0 \\ 0 & b \end{bmatrix} \quad \text{is p.d.}$$

or, $a > 0$ and $b > 0$.

Since n is even, we may use Theorem 6.10,

$$C = \begin{bmatrix} a(b^2 + b + a^2) & a^2 \\ a^2 & a(1+b) \end{bmatrix} \quad \text{is p.d.}$$

which implies

$$\text{(i)}\quad a(b^2 + b + a^2) > 0$$
$$\text{(ii)}\quad \left((b+1)^2 + a^2\right)a^2 b > 0$$

or, $a > 0$ and $b > 0$, as above.

We conclude that in the special case of the left half plane, our result, although more complicated, is equivalent to the Hermite form. To fully understand the reason, recall that Hermitian form $H = [h_{ij}]$ can be constructed via Bezoutian form

$$\frac{\bar{\Delta}(\lambda)\Delta(s) - \Delta(-\lambda)\bar{\Delta}(-s)}{\lambda + s} = \sum_{i,j=1}^{n} h_{ij}\lambda^{i-1}s^{j-1}. \tag{6.32}$$

Note that the quotient in (6.32) is a polynomial since $(\lambda + s)$ is a factor in the numerator. On the other hand, (6.23), (6.24) in our approach, imply

$$\frac{-q_0 \sum_{k=1}^{n} (\lambda s)^{k-1}}{\phi(\lambda, s)} = \sum_{k,m=1}^{n} c_{km} \lambda^{k-1} s^{m-1} \tag{6.33}$$

$$\frac{-q_0 \sum_{k=1}^{n} (\lambda s)^{k-1}}{\lambda + s} = \sum_{k,m=1}^{n} c_{km} \lambda^{k-1} s^{m-1} \tag{6.34}$$

only on the spectrum of $\Delta(\lambda)$; that is, only for $\lambda = \lambda_i$ $s = \bar{\lambda}_j$, the roots of $\Delta(\lambda)$. The quotient in (6.33) is not a polynomial since $\phi(\lambda, s)$ is ***not a factor*** in the numerator. One may try to improve on our results by searching for a better quotient such that the right hand side is a polynomial. However we were not able to do so for the M-transformable family. We conclude that in the expense of the criterion's complexity we are able to construct a criterion for a large family of regions. On the other hand, if we wish a divisible quotient for the family of M-transformable regions, the complexity increases. This is done using a generalized Bezoutian form in the next section.

Generalized Bezoutian Form

For the left half plane, we have mentioned that the Bezoutian (6.32) leads to Hermite criterion. Likewise, for the unit disk, Schur-Cohn matrix $S = [s_{ij}]$ can be constructed via Bezoutian form

$$\frac{\Delta(\lambda)\bar{\Delta}(s) - \lambda^n \bar{\Delta}(\frac{1}{\lambda}) s^n \Delta(\frac{1}{s})}{-1 + \lambda s} = \sum_{i,j=1}^{n} s_{ij} \lambda^{i-1} s^{j-1} . \tag{6.35}$$

We now present a general Bezoutian form. Let $\aleph$ be defined according to (6.1)–(6.3), and consider the n–th order polynomial $\Delta(\lambda)$ with roots $\lambda_1, \lambda_2, \ldots, \lambda_n$. Denote $m = \deg_\lambda \phi(\lambda, s) = \deg_s \phi(\lambda, s)$. For fixed λ, $\omega_i(\lambda)$, $i = 1, 2, \ldots, m$, are all functions satisfying $\phi(\lambda, \omega_i(\lambda)) = 0$. Likewise, for fixed s, $z_j(s)$, $j = 1, 2, \ldots, m$, are all functions satisfying $\phi(z_j(s), s) = 0$. Since $\phi(\lambda, s)$ is Hermitian, we can arrange i, j such that $\omega_i(\lambda) = \overline{z_j(\bar{\lambda})}$.
Now, define

$$B(\lambda, s) = \frac{\prod_{i,j} \left[\Delta(\lambda)\bar{\Delta}(s) - \Delta(z_i(s))\bar{\Delta}(\omega_j(\lambda))\right]}{-\phi(\lambda, s)} . \tag{6.36}$$

Next, we present the following results.

Lemma 6.17 (i) Suppose $\phi(\lambda,s)$ satisfies $\phi_{mj} = 0 \ \forall\, j \geq 1$. Then the numerator in (6.36) is a polynomial in λ, s. (ii) If $\phi(\lambda,s)$ is an irreducible polynomial, or a product of prime irreducible polynomials, then $B(\lambda,s)$ in (6.36) is a polynomial in λ, s.

Proof We first show that the numerator in (6.36) is, in general, a rational function. To this end, consider

$$b(\mu; \lambda, s, w) = \mathrm{Res}_z \left[\mu - (\Delta(\lambda)\, \bar{\Delta}(s) - \Delta(z)\, \bar{\Delta}(w)),\ \phi(z, s)\right].$$

Thus,

$$b(\mu; \lambda, s, w) = 0 \Leftrightarrow \exists\, i \ \text{ such that } \ \mu = \Delta(\lambda)\, \bar{\Delta}(s) - \Delta(z_i(s))\, \bar{\Delta}(w).$$

Now consider

$$c(\mu; \lambda, s) = \mathrm{Re\,s}_w[\, b(\mu; \lambda, s, w),\quad \phi(\lambda, w)\,]\,.$$

Thus,

$$c(\mu; \lambda, s) = 0 \Leftrightarrow \exists\, j \quad \text{such that} \quad \mu = \Delta(\lambda)\, \bar{\Delta}(s) - \Delta(z_i(s))\, \bar{\Delta}(w_j(\lambda))\,.$$

Denote by $d(\lambda,s)$ the leading coefficient in μ in the polynomial $c(\mu; \lambda,s)$. Then,

$$[\text{numerator of } (6.36)] = \frac{c(0; \lambda, s)}{d(\lambda, s)}\,. \tag{6.36'}$$

Next, to show that (6.36') is indeed a polynomial, recall that a continuous rational function in $\mathbf{C}^2$ is a polynomial. Thus, it is sufficient to show that the numerator in (6.36) is continuous in λ and s, or that every factor is continuous. In other words, we have to show that $z_i(s)$ is continuous (this implies, by symmetry, that $w_j(\lambda)$ is continuous). Since, by hypothesis, the leading coefficient of $\phi(\lambda,s)$ as a polynomial in s is constant, it follows that it is monic. However, $z_i(s)$ is a zero of $\phi(z;s)$, a monic polynomial in z with polynomials in s as coefficients. Thus $z_i(s)$ is continuous. This proves part (i) of the Lemma. To prove (ii), note that $\phi(\lambda,s) = 0$ implies that the numerator in (6.36) vanishes.
To illustrate $\phi_{mj} = 0 \ \ \forall j \geq 1$, consider the ellipse

$$\left\{x + iy: \frac{x^2}{a^2} + \frac{y^2}{b^2} - 1 < 0\right\}\,.$$

Here,

$$\phi(\lambda, s) = \lambda^2 (b^2 - a^2) + 2\lambda s (b^2 + a^2) + s^2 (b^2 - a^2) - 4a^2 b^2\,.$$

We see that if $a \neq b$, then $m = 2$ and $\phi_{mj} = 0 \ \forall_j \geq 1$. However, if $a = b$ (the unit disk), then $m = 1$ and $\phi_{11} \neq 0$ so that part (i) does not hold. In other words, all ellipses satisfy condition (i) except the unit disk, for which (6.35) serves as a Bezoutian form. Note that this limitation does not exist in Theorem 2. We are now in a position to define a Hermitian matrix C.

$$q_o(-1)^{m^2}B(\lambda, s)\sum_{i=1}^{n}(\lambda s)^{i-1} \overset{\text{mod}}{=} \sum_{i,j=1}^{n} c_{ij}\lambda^{i-1}s^{j-1} \qquad ; \quad C=[c_{ij}] \tag{6.37}$$

where $\overset{\text{mod}}{=}$ stands for "λ, s modulo $\Delta(\cdot)$, $\bar{\Delta}(\cdot)$", respectively.
Define

$$a(\lambda) = \prod_j \overline{\Delta(z_j(\bar{\lambda}))} \quad ; \quad a_{km} = a(\lambda_k)\,\overline{a(\lambda_m)} \quad ; \quad A = [a_{km}]\,. \tag{6.38}$$

Then, on the roots $\{\lambda_k, \lambda_m\}$, the following holds

$$q_oA \cdot TT^* = [-\phi(\lambda_k, \bar{\lambda}_m)] \cdot TCT^*\,. \tag{6.39}$$

Theorem 6.11 Let $\Delta(\lambda) = \sum_{i=o}^{n} a_i\lambda^i$ and consider an M–transformable region $\aleph$. Let C be defined by (6.37). Then $\sigma(\Delta) \subset \aleph$, if and only if q_oC is p.d.

To complete our construction, we describe a simple way to generate matrix C.

Define the polynomial

$$d(x;\lambda) = \mathrm{Res}_v[x - \Delta(v), \phi(v, \lambda)] \tag{6.40}$$

and denote by $d_R(\lambda)$ the leading coefficient of $d(x;\lambda)$ as a polynomial in x. Then

$$\bar{a}(\lambda) = (-1)^R \frac{d(o;\lambda)}{d_R(\lambda)}\,. \tag{6.41}$$

Finally, $C = [c_{km}]$ is obtained via

$$a(\lambda)\bar{a}(s)\sum_{m=1}^{n^2} q_m\phi^{m-1}(\lambda, s)\sum_{k=1}^{n}(\lambda, s)^{k-1} \overset{\text{mod}}{=} \sum_{k,m=1}^{n} c_{km}\lambda^{k-1}s^{m-1}\,. \tag{6.42}$$

To compare with matrix C in the above **Discussion,** let $\Delta(\lambda) = \lambda^2 + a\lambda + b$. Then for the left half plane,

$$C = 8a^3b\begin{bmatrix} a^2+b+1 & ab \\ ab & b(b+1) \end{bmatrix} = 2q_ob\begin{bmatrix} a^2+b+1 & ab \\ ab & b(b+1) \end{bmatrix}.$$

Finally we comment that from a computational point of view, Theorem 6.9 is superior to Theorem 6.11 (compare (6.40) to (6.22)). The latter was presented merely to show the connection to generalized Bezoutians and divisible quotients.

Region Intersection

In some applications we need a root clustering criterion with respect to region intersection. For example, the left hyperbola, the intersection of a double hyperbola and a half plane, is important for relative stablility of linear dynamical systems.

Fact Let $\Delta(\lambda) = \sum_{i=0}^{n} a_i \lambda^i$, $\aleph = \aleph_1 \cap \aleph_2$, and suppose

$\sigma(\Delta) \subset \aleph_1 \Leftrightarrow \{\Delta_{1i} > 0\}$ and $\sigma(\Delta) \subset \aleph_2 \Leftrightarrow \{\Delta_{2i} > 0\}$. Then,

$\sigma(\Delta) \subset \aleph \Leftrightarrow \{\Delta_{1i} > 0\} \cup \{\Delta_{2i} > 0\}$.

In other words, if two regions have root clustering criteria, their intersection has the union of the respective criteria as a root clustering criterion. But what about the case where one of the members has no criteria ? A simple extension of the previous results shows the following criterion.

Theorem 6.12 Let $\Delta(\lambda) = \sum_{i=0}^{n} a_i \lambda^i$, $\aleph = \aleph_1 \cap \aleph_2$, and suppose $\aleph$ and $\aleph_2$ are M–transformable. Suppose also that $[q_o C]_{\aleph_1}$ and $[q_o C]_{\aleph_2}$ are generated by (6.24) for $\aleph_1$ and $\aleph_2$ respectively. Then $\sigma(\Delta) \subset \aleph$, if and only if

(i) $[q_o C]_{\aleph_1}$ is p. d.,

(ii) $[q_o C]_{\aleph_2}$ is p. d..

In other words, for the family of M-transformable regions, it is not necessary for both members of the intersection to be transformable. However, we do require that one member as well as the intersection be transformable.

Finally, we extend our results as follows. Let $\eta(\lambda,s)$ be Hermitian, and define $C = [d_{km}]$ via

$$\eta(\lambda, s) \sum_{m=1}^{n^2} q_m \phi^{m-1}(\lambda, s) \sum_{k=1}^{n} (\lambda s)^{k-1} \overset{\text{mod}}{=} \sum_{k,\, m=1}^{n} c_{km} \lambda^{k-1} s^{m-1} \tag{6.43}$$

Then, similar to our previous results, matrix C satitisfies

$$q_o[\eta(\lambda_i, \bar{\lambda}_j)] \cdot TT^* = [-\phi(\lambda_i, \bar{\lambda}_j)] \cdot TCT^*$$

or

$$q_o \left[\frac{\eta(\lambda_i, \bar{\lambda}_j)}{-\phi(\lambda_i, \bar{\lambda}_j)} \right] \cdot TT^* = TCT^* \tag{6.44}$$

Now define

$$\aleph = \{\lambda \in C: \phi(\lambda, \bar{\lambda}) < 0\} \cap \{\lambda \in C: \zeta(\lambda, \bar{\lambda}) > 0\}$$

$$\stackrel{\Delta}{=} \aleph_1 \cap \aleph_2 \text{ such that } \zeta(\lambda, \bar{\lambda}) > 0 \Rightarrow \eta(\lambda, \bar{\lambda}) > 0 . \tag{6.45}$$

Definition 6.2 Region $\aleph$ defined by (6.45) is F-transformable, if all $\alpha_i \in \aleph$ imply

$$F(\alpha_1, \dots, \alpha_n) \stackrel{\Delta}{=} \left[\frac{\eta(\alpha_i, \bar{\alpha}_j)}{-\phi(\alpha_i, \bar{\alpha}_j)} \right]_{i, j = 1, \dots, n} \quad \text{is p.d..}$$

Theorem 6.13 Let $\Delta(\lambda) = \sum_{i=1}^{n} a_i \lambda^i$ and $\aleph = \aleph_1 \cap \aleph_2$ be given by (6. 45).

Suppose $\aleph$ is F-transformable and $\aleph_2$ is M-transformable.

Then, $\sigma(\Delta) \subset \aleph$, if and only if

(i) $[q_0 C]_\aleph$, generated by (6.43), is p.d.

(ii) $[q_0 C]_{\aleph_2}$, generated by (6.24), is p.d..

NOTES AND REFERENCES

The matrix equation approach has a long history. As noted in Chapter 2, Theorem 2.8 is traced back to 1982, due to Lyapunov [1]. In 1952, Stein [1] developed Theorem 2.9. These two classical results deal with the left half plane and the unit disk. A more general case, Rank(Φ) - Sign(Φ) = 2, given in Theorem 6.1, was first obtained by Schneider [1] in 1963. The proof, however, is due to Kharitonov [1]. Lemma 6.1 is due to Schur [2]. A proof to a weaker version of the Lemma can be found in Bellman [1]. The proof we present was communicated to us by J.M. Exbrayat. M-Transformability and Theorem 6.5 are due to Mazko [1]. We know that M-transformability includes Rank (Φ) - Sign(Φ) = 2, but the former is almost impossible to be checked. In addition to Theorem 6.5, Mazko states that our Theorem 6.3 (a), holds for any, not just simple, matrix A. Our investigation does not support his conclusion. In the proof of Lemma 6.5 we use results on Isom ($\mathcal{H}_n$) from Cartan [1]. Section 6.5 is taken from Gutman and Taub [1], and Section 6.6 from Taub and Gutman [2]. Lemma 6.16 is due to Oppenheim [1].

Chapter 7 : PARAMETER SPACE AND FEEDBACK DESIGN

So far we have answered the question: given a polynomial (or a matrix), find necessary and sufficient conditions such that all the zeros (eigenvalues) of the polynomial (matrix) lie in a given region in the complex plane. However, in feedback design, we are interested in the inverse problem. That is, given a linear plant, we are looking for a compensator such that the closed loop poles lie in a prescribed region in the complex plane. Such a design results in ***relative stability*** of the closed loop. Moreover, to avoid ***inverse response*** we may impose an additional requirement; namely, the compensator's zeros should lie in a prescribed region. Yet, another possible requirement is a ***stable compensator***. The aim of the present chapter is to show how to achieve such requirements with a fixed (or minimal) order compensator. We will describe both polynomial and matrix versions and point out the advantages of each. However, before discussing feedback design we describe some important properties of the space spanned by the parameters of the system. Our main concept in the parameter space is the ***critical constraint***. We show that the boundary of the root clustering region in the parameter space is defined by ***a single*** inequality, the critical constraint. This inequality contains, in addition to the root clustering region, some additional branches, which have to be eliminated. The elimination is obtained by solving a set of polynomial equations.

7.1 CONCEPTS

Following our previous results, given $f \in R[x,y]$ we define an algebraic region

$$\aleph = \{x + iy : f(x,y) < 0\} . \tag{7.1}$$

Using

$$\phi(\alpha, \beta) = f\left(\frac{\alpha+\beta}{2}, \frac{\alpha-\beta}{2i}\right) = \sum_{i,j} \phi_{ij}\, \alpha^i \beta^j \tag{7.2}$$

we also have

$$\aleph = \{\lambda \in C : \phi(\lambda,\bar{\lambda}) < 0\}. \tag{7.3}$$

Given a matrix $A(p) \in C^{n \times n}[p_1, p_2, \ldots, p_m]$ with spectrum $\sigma(A)$, or a polynomial

$$\Delta(\lambda; p) = \sum_{i=0}^{n} a_i(p)\lambda^i \quad , \qquad a_i(p) \in C[p_1, p_2, \ldots, p_m] \tag{7.4}$$

with spectrum (vanishing set) $\sigma(\Delta)$, we define

$$\text{(i)} \quad \tilde{\aleph} = \{p \in \mathbf{R}^m : \ \sigma(A(p)) \subset \aleph\}$$

or (7.5)

$$\text{(ii)} \quad \tilde{\aleph} = \{p \in \mathbf{R}^m : \ \sigma(\Delta(\lambda; p)) \subset \aleph\} .$$

We say that $\tilde{\aleph}$ is the ***stability region in the parameter space*** R^m.
As an illustration, if $\aleph$ admits a root clustering criterion, then there exists a set of polynomial inequalities in p defining $\aleph$.

$$\tilde{\aleph} = \{p \in \mathbf{R}^m : \; q_i(p) > 0 , \quad i = 1, 2, \ldots, L\}. \tag{7.6}$$

Example 7.1 Consider the characteristic polynomial

$$\Delta(\lambda) = \lambda^3 + p_1\lambda^2 + (p_2 - 5p_1 - 13)\lambda + p_2$$

and let $\aleph$ be the left half plane. Using Routh-Hurwitz criterion, we find

$$q_1 = p_1 > 0$$

$$\text{(i)} \qquad q_2 = p_1(p_2 - 5p_1 - 13) - p_2 > 0$$

$$q_3 = p_2 q_2 > 0$$

or, in a different form (Lienard-Chipart)

$$q_1 = p_1 > 0$$

$$\text{(ii)} \qquad q_2 = p_1(p_2 - 5p_1 - 13) - p_2 > 0$$

$$q_3 = p_2 > 0 \; .$$

Thus, in the parameter space R^2, the stability region $\aleph$ is connected.
The following example shows that this is not always the case.

Example 7.2 Consider the characteristic polynomial

$$\Delta(\lambda) = \lambda^3 - (2.1 - p_2)\lambda^2 + 2\lambda - (0.6 - p_1)$$

and let $\aleph$ be the unit disk. Using Schur-Cohn-Jury criterion, we find

$$q_1 = p_1 + p_2 + 0.3 > 0$$

$$q_2 = - p_1 - p_2 + 5.7 > 0$$

$$q_3 = 3 - (0.6 - p_1)(2.7 - p_1 - p_2) > 0$$

$$q_4 = - 1 + (p_1 - 0.6)(- p_1 + p_2 - 1.5) > 0 .$$

The region $\tilde{\aleph}$ given by (7. 6) and the above inequalities is disconnected. It consists of two cells.
The region $\tilde{\aleph}$ defined in (7. 6) can be written as

$$\tilde{\aleph} = \bigcap_{i=1}^{L} \{p \in \mathbf{R}^m : q_i(p) > 0\}\,. \tag{7.7}$$

Investigating this intersection for various examples, one finds that along the boundary $\partial\tilde{\aleph}$, not all the polynomials $q_i(p)$ vanish. In the next section we elaborate on this point.

7.2 THE CRITICAL CONSTRAINTS

Given $A(p) \in C^{n\times n}[p_1,p_2,\ldots,p_m]$, we set

$$\text{(i)} \qquad c(p) = (-1)^n \,|\sum \phi_{ij} A^i \otimes \bar{A}^j| \tag{7.8}$$

Given a polynomial $\Delta(\lambda; p) = \sum_{i=0}^{n} a_i(p)\lambda^i$ with $a_i(p) \in C[p_1, p_2, \ldots, p_m]$, we set

$$\text{(ii)} \qquad c(p) = \mathrm{Res}[\Delta(\lambda; p),\ \mathrm{Res}[\Delta(s; p),\ -\phi(\lambda, s)]] \tag{7.8'}$$

where Res is the Resultant of two polynomials.

Definition 7.1 The polynomial c(p) is called the ***critical constraint.***

To fully understand the properties of c(p), recall the polynomial $q(\eta)$ whose roots are $\phi(\lambda_i, \bar{\lambda}_j)$, $i, j = 1, 2, \ldots, n$, where λ_i are the eigenvalues (roots) of a given matrix A (polynomial Δ). That is

$$q(\eta) = \prod_{i,j=1}^{n} (\eta - \phi(\lambda_i, \bar{\lambda}_j))\,. \tag{7.9}$$

Using Theorem 5.5, given $A \in C^{n\times n}$,

$$q(\eta) = |\eta I - \phi(A \otimes \bar{A})| \tag{7.10}$$

where

$$\phi(A \otimes \bar{A}) = \sum \phi_{ij} A^i \otimes \bar{A}^j\,.$$

Using Theorem 5.4 given $\Delta(\lambda) \in C[\lambda]$,

$$q(\eta) = \mathrm{Res}[\Delta(\lambda),\ \mathrm{Res}[\Delta(s),\ \eta-\phi(\lambda,s)]] \tag{7.11}$$

Thus, c(p) is the free coefficient of the polynomial $q(\eta)$. In other words, using (7.9)

$$c(p) = \prod_{i,j=1}^{n} (-\phi(\lambda_i, \bar{\lambda}_j))$$

$$= \left[(-1)^n \prod_{i=1}^{n} \phi(\lambda_i, \bar{\lambda}_i)\right]\left[\prod_{i<j} |\phi(\lambda_i, \bar{\lambda}_j)|^2\right]. \qquad (7.12)$$

Now, if $x + iy \in \partial\aleph$, then $\phi(\lambda, \bar{\lambda}) = f(x, y) = 0$, since $\partial\aleph \subset V(f)$, the vanishing set of f. Thus, using (7.12), the boundary $\partial\tilde{\aleph}$ of the stability region $\tilde{\aleph}$ satisfies

Fact 7.1 $\quad \partial\tilde{\aleph} \subset V(c) \stackrel{\Delta}{=} [p \in \mathbf{R}^m : c(p) = 0\}. \qquad (7.13)$

This relation explains the importance of the critical constraint.

Definition 7.2 f is H-transformable if $\phi(\alpha, \bar{\beta}) \neq 0$ for all $\alpha, \beta \in \aleph$.

Clearly, any f discussed in Chapters 5 and 6 is H-transformable. As a direct consequence of (7.12), we also have the following fact.

Fact 7.2 If f is H-transformable, then

$$\tilde{\aleph} \subset \Omega \stackrel{\Delta}{=} \{p \in \mathbf{R}^m : c(p) > 0\}. \qquad (7.14)$$

Otherwise, we still have

$$\tilde{\aleph} \subset [p \in \mathbf{R}^m : c(p) \geq 0\}. \qquad (7.14')$$

Based on Theorem 5.18, in the case of a real matrix A and a symmetric region $\aleph$, c(p) can be factored as $c(p) = c_1(p)\, c_2(p)$.

Fact 7.3 Let $A \in R^{n \times n}$ and suppose $\aleph$ is symmetric with respect to the real axis. Then $c(p) = c_1(p)\, c_2(p)$, where

$$c_1(p) = \det[-\sum_i f_{i0} A^i]$$

$$c_2(p) = \det[-\phi(A \odot A)] \qquad (7.15)$$

where $\odot$ is the bialternate product.

As an illustration, if $\Delta(\lambda) = \sum_{i=0}^{n} a_i \lambda^i$ and $\aleph$ is the **left half plane**,

$$c(p) = |H_n(p)| = a_0 |H_{n-1}(p)| \tag{7.16}$$

where H_n is the Hurwitz determinant of order n

$$H_n = \begin{bmatrix} a_{n-1} & a_{n-3} & a_{n-5} & \cdots \\ a_n & a_{n-2} & a_{n-4} & \cdots \\ 0 & a_{n-1} & a_{n-3} & \cdots \\ 0 & a_n & a_{n-2} & \cdots \\ \vdots & & & \end{bmatrix}. \tag{7.17}$$

If $\aleph$ is the unit disk,

$$c(p) = (-1)^n \Delta(1)\Delta(-1) | \Delta^-_{n-1}(p)| \tag{7.18}$$

where $\Delta^-_{n-1} = X_{n-1} - Y_{n-1}$,

$$X_{n-1} = \begin{bmatrix} a_n & a_{n-1} \cdots & a_2 \\ & a_n \cdots & a_3 \\ \bigcirc & & \vdots \\ & & a_n \end{bmatrix} \qquad Y_{n-1} = \begin{bmatrix} & & & a_0 \\ & \bigcirc & & \vdots \\ & & a_0 & \cdots a_{n-3} \\ a_0 & & a_1 & \cdots a_{n-2} \end{bmatrix}. \tag{7.19}$$

One may try a different approach to the critical constraint.

Set $\lambda = x + iy$ in $\Delta(\lambda) = \sum_{i=0}^{n} a_i \lambda^i$ and expand. Then (see section 3.2),

$$\Delta(\lambda) = \Delta_r(x, y) + i\Delta_i(x, y). \tag{7.20}$$

Thus $\lambda \in \partial\aleph \cap \sigma(\Delta)$ implies

$$\Delta_r(x, y) = 0, \qquad \Delta_i(x, y) = 0, \qquad f(x, y) = 0. \tag{7.21}$$

Let

$$\begin{aligned} \Delta_1(y) &= \mathrm{Res}_x[f(x, y), \ \Delta_r(x, y)] \\ \Delta_2(y) &= \mathrm{Res}_x[f(x, y), \ \Delta_i(x, y)]. \end{aligned} \tag{7.22}$$

Then, we may define the critical constraint as follows.

$$c(p) = \mathrm{Res}[\Delta_1(y),\ \Delta_2(y)]\ . \qquad (7.23)$$

As an illustration, let $\aleph$ be the left half plane. Since $\partial\aleph$ is defined by $x = 0$, we obtain

$$c(p) = \mathrm{Res}[\Delta_r(0,y)\ ,\ \Delta_i(0,y)]\ . \qquad (7.24)$$

This $c(p)$ vanishes as soon as a zero of $\Delta(\lambda)$ appears on the boundary. However, this approach suffers two difficulties. First, we can not verify that (7.23) is sign invariant in $\aleph$, as we have in (7.14). Second, (7.23) is computationally more complicated than our previous approach.
It is important to note that all the properties of the critical constraint hold for any algebraic region. However, Fact 7.1 reveals that $c(p) = 0$ only partially characterizes $\partial\aleph$. If we wish a full characterization we have to restrict the region $\aleph$. Recall the root clustering results of Chapter 5.

Theorem 7.1 Suppose $\aleph$ is IR-transformable and simple, with $\{q_i\}$ as the root clustering coefficients. Then

(i) $\tilde{\aleph} = \{p \in R^m : q_L(p) > 0 ,\dots, q_1(p) > 0,\ c(p) > 0\}$

(ii) $cl(\tilde{\aleph}) = \{p \in R^m : q_L(p) \geq 0 ,\dots, q_1(p) \geq 0\ , c(p) \geq 0\}$, the closure of $\aleph$ (7.25)

(iii) $\partial\tilde{\aleph} = \{p \in R^m : q_L(p) \geq 0 ,\dots, q_1(p) \geq 0\ ,\ c(p) = 0\}$.

Proof From root clustering theory, (i) holds for transformable regions. If $\aleph$ is also simple, then (ii) holds. To prove (iii), note that $p \in \partial\tilde{\aleph} \Rightarrow p \in cl(\tilde{\aleph})$ and together with Fact 1, $p \in$ RHS of (iii). Conversely, if $p \in$ RHS of (iii), then $p \in cl(\tilde{\aleph})$, but according to (i), $p \notin \tilde{\aleph}$. Thus $p \in \partial\tilde{\aleph}$. This completes the proof.

As an illustration, recall that for P-transformable regions, $\{q_i\}$ are the coefficients of the polynomial (7.10), and in the real symmetric case it reduces to the coefficients of the polynomials $|\eta I - \sum_i f_{i0} A^i|$ and $|\eta I - \phi(A \odot A)|$.

Remark 7.1 Conditions (7.25), (ii) - (iii) do not hold for any root clustering criterion.
As an illustration, consider Example 7.1 . In this example,

$$\tilde{\aleph} = \{(p_1, p_2): q_1 > 0,\ q_2 > 0,\ q_3 > 0\}.$$

However, although in version (i), $q_3(p) = c(p)$, we have

$$\partial\tilde{\aleph} \neq \{(p_1, p_2): q_1 \geq 0,\ q_2 \geq 0,\ q_3 = 0).$$

This is so, since $p_1 = p_2 = 0$ implies $q_1 = q_2 = q_3 = 0$, but $0 \notin \partial\aleph$.

On the other hand, using composite real matrices, $\sigma(A)$ lies in the left half plane, if and only if

(i) Coef $|\eta I - A| > 0$,

(ii) Coef $|\eta I - (A \odot I + I \odot A)| > 0$.

If we take A as the companion matrix of $\Delta(\lambda)$ in Example 7.1, we find

$$A \odot I + I \odot A = \frac{1}{2}\begin{bmatrix} 0 & 1 & 0 \\ -p_2 + 5p_1 + 13 & -p_2 & 1 \\ p_2 & 0 & -p_1 \end{bmatrix}$$

so that

$$q_1 = p_1 > 0$$

$$q_2 = p_2 - 5p_1 - 13 > 0$$

$$q_3 = p_2 > 0$$

$$q_4 = 2p_2 > 0$$

$$q_5 = p_1^2 + p_2 - 5p_1 - 13 > 0$$

$$q_6 = p_1(p_2 - 5p_1 - 13) - p_2 > 0$$

and

$$c(p) = q_3 q_6 .$$

Now it is clear that

$$\partial\tilde{\aleph} = \{(p_1, p_2): q_1 \geq 0, q_2 \geq 0, q_4 \geq 0, q_5 \geq 0, c(p) = 0\}$$

and in fact

$$\partial\tilde{\aleph} = \{(p_1, p_2): q_2(p) \geq 0,\ c(p) = 0\} .$$

Note also that $p_1 = p_2 = 0$ does not imply $q_i = 0 \ \forall\ i$. Indeed, $0 \notin \partial\tilde{\aleph}$.

We see that the root clustering criterion based on composite matrices has an important advantage in the parameter space in spite of n^2 or $\frac{1}{2}\,n(n+1)$ inequalities rather than the minimum, n.

7.3 ADMISSIBLE POINTS IN THE PARAMETER SPACE

The purpose of this section is two fold:

$$\begin{aligned} &\text{(i)} \quad \text{Is } \tilde{\aleph} = \emptyset, \text{ the empty set?} \\ &\text{(ii)} \quad \text{If } \tilde{\aleph} \neq \emptyset, \text{ generate points in } \tilde{\aleph}. \end{aligned} \tag{7.26}$$

It is important to note that solving the above ***algebraically*** is not a simple task. First, we transform the (root clustering) inequalities $\{q_i(p) > 0\}$ into equalities and solve the minimization problem.

$$\begin{aligned} &\text{Minimize} \qquad J = \sum_{k=1}^{\infty} (p_k - \gamma_k)^2 + \sum_{i=1}^{L} (t_i - \tau_i)^2 \\ &\text{subject to} \qquad t_i^2 q_i(p) - 1 = 0 \qquad i = 1, \ldots, L, \end{aligned} \tag{7.27}$$

where $\{t_i\}$ are additional variables, and $\{\gamma_k\}$, $\{\tau_i\}$ are prespecified constants. Using the method of Lagrange multipliers, we have

$$\begin{aligned} &t_i^2 q_i(p) - 1 = 0, \qquad i = 1, 2, \ldots, L \\ &2(p_k - \gamma_k) - \sum_{j=1}^{L} \frac{\partial q_j(p)}{\partial p_k} (t_j - \tau_j) t_j^3 = 0, \quad k = 1, 2, \ldots, m. \end{aligned} \tag{7.28}$$

Theorem 7.2 Consider the set (7.28) of L+m equations with L+m unknowns (the p_k's, and the t_i's) and suppose (i) $q_i(p)$ contains no multiple factors, and (ii) $q_i(p)] \neq [r(p)]^2 \prod_{j \in S, j \neq i} q_j(p)$, where $r(p) \in \mathbf{R}(p)$, the field of real rational functions of p, and $S \subset \{1,2,\ldots,L\}$. Then, (7.28) has a finite number of solutions in C^{L+m} for almost all $(\gamma_1,\ldots,\gamma_m, \tau_1,\ldots,\tau_L) \in R^{L+m}$.

It can be shown that conditions (i) and (ii) do not cause significant loss of generality. Now, choosing $\{\gamma_k\}$ and $\{\tau_i\}$ randomly, we calculate all the real solutions of (7.28). These real solutions are clearly in $\tilde{\aleph}$. If there exists no real solution to (7. 28), we conclude that $\tilde{\aleph}$ is empty. One way to solve (7.28) is the use of the Resultant method for elimination. However , in many cases, solving L+m equations (using elimination) calls for high computer memory and computing time. Yet, we wish to take advantage of the fact that in Theorem 7.2 the number of solutions is ***finite***. Note that ***without this important property any method is useless***. Thus, we combine Theorem 7.2 with our results on the critical constraint to reduce the computation complexity of (7.28). As we have already seen

$$\tilde{\aleph} \subset \{p \in \mathbf{R}^m : c(p) > 0\}$$
$$\Omega = \{p \in \mathbf{R}^m : c(p) > 0\} . \qquad (7.29)$$

Let Ω_i be the connected components of Ω. That is,

$$\Omega = \bigcup_{i \in I} \Omega_i \qquad (7.30)$$

where I is a ***finite*** (see Section 7.5 for proof) index set. We propose to solve (7.26) in two basic steps. First we generate at least one point at each Ω_i. Then, we use the root clustering inequality set to test whether such a point lies in $\tilde{\aleph}$. Towards this end, consider the minimization problem.

$$\text{Minimize} \qquad J = \sum_{k=1}^{m} (p_k - \gamma_k)^2 + (t - \tau)^2$$
$$\text{subject to} \qquad t^2 c(p) - 1 = 0 . \qquad (7.31)$$

Using the method of Lagrange multipliers, we have

$$t^2 c(p) - 1 = 0$$
$$2(p_k - \gamma_k) - (t - \tau)\, t^3 \frac{\partial c(p)}{\partial p_k} = 0 , \qquad k = 1, 2, \ldots, m . \qquad (7.32)$$

The solution to (7.26) is obtained using the following algorithm.

Algorithm 7.1 (Admissible points in $\tilde{\aleph}$)

Data: (i) A polynomial $\Delta(\lambda; p)$ or a matrix $A(p)$,

(ii) A transformable region $\aleph \subset \mathbf{C}$, with f H-transformable.

Step 1: Calculate c(p), recalling (7.8), or (7.15) - (7.19) and factorize c(p) in $\mathbf{R}[p_1,\ldots,p_m]$.

Step 2: If c(p) has factors of odd multiplicity, reduce all of them to first order.

Step 3: If c(p) has factors of even multiplicity, reduce all of them to first order, and in what follows consider two critical constraints: c(p) and -c(p).

Step 4: Choose $\{\gamma_k\}$ and τ randomly. Solve the set (7.32) of m+1 equations with m+1 unkowns. This can be done algebraically using the ***resultant*** method, or numerically using the ***continuation*** method. Save all real solutions $\{p^s\}$.

Step 5: Test each real solution p^s for $p^s \in \tilde{\aleph}$ (either numerically, or) by checking **all** the root clustering inequalities $q_i(p^s) > 0$, $i = 1,2,\ldots,L$. If no such points exist, we conclude that $\aleph$ is empty. However, should such points in fact exist, they should be saved, as these are the required points. In particular in each component of $\tilde{\aleph}$ we find at least one admissible point.

7.4 COMPENSATOR DESIGN — POLYNOMIAL FORM

Consider the SISO feedback configuration shown in Figure 7.1, where

$$G(s) = n(s)/d(s), \quad C(s) = a(s) / b(s) \tag{7.33}$$

and

$$a(s) = \sum_i a_i s^i, \quad b(s) = \sum_i b_i s^i, \quad \deg(b) \geq \deg(a), \tag{7.34}$$

G(s) is the plant transfer function, given to the designer, while the compensator (controller) transfer function C(s) has to be selected by the designer.

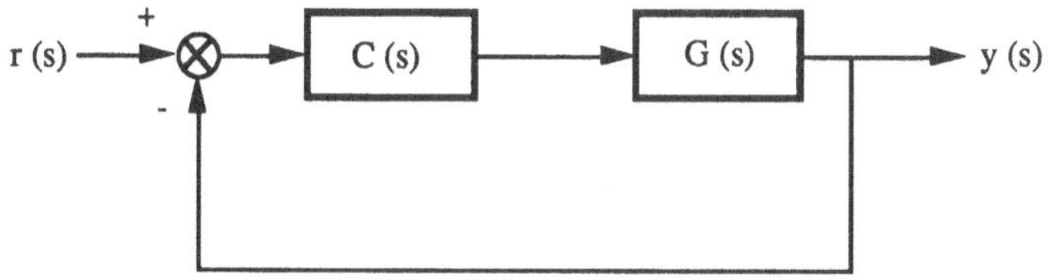

Figure 7.1 Unit feedback SISO system.

The closed loop characteristic polynomial is given by

$$\Delta(s) = n(s)\, a(s) + d(s)\, b(s) \ . \tag{7.35}$$

For MIMO feedback configuration, let

$$G(s) = n(s)d^{-1}(s), \quad C(s) = b^{-1}(s)a(s) \tag{7.36}$$

where now, n(s), d(s), a(s) and b(s) are polynomial matrices of proper dimensions. The closed loop characteristic polynomial becomes

$$\Delta(s) = |a(s)\, n(s) + b(s)d(s)| \ . \tag{7.37}$$

In the state space, let the plant be

$$\dot{x}_p = A_p x_p + B_p u \ , \qquad y = C_p x_p \tag{7.38}$$

where

$$x_p \in \mathbf{R}^{n_p}, \quad u \in \mathbf{R}^{m_p}, \quad y \in \mathbf{R}^{r_p}, \quad A_p \in \mathbf{R}^{n_p \times n_p}, \quad B_p \in \mathbf{R}^{n_p \times m_p}, \quad C_p \in \mathbf{R}^{r_p \times n_p}.$$

To this system we apply the linear feedback

$$u = Fy + r. \tag{7.39}$$

The closed loop system becomes

$$\dot{x}_p = (A_p + B_p F C_p) x_p + B_p r. \tag{7.40}$$

The closed loop characteristic polynomial is

$$\Delta(s) = |sI - (A_p + B_p F C_p)|. \tag{7.41}$$

Likewise, it is possible to construct a ***dynamic*** feedback compensator of fixed order n_c.

$$\begin{aligned} \dot{x}_c &= A_c x_c + B_c (r + y) \\ u &= C_c x_c + D_c y \end{aligned} \tag{7.42}$$

where

$$x_c \in \mathbf{R}^{n_c}, \quad A_c \in \mathbf{R}^{n_c \times n_c}, \quad B_c \in \mathbf{R}^{n_c \times r_p}, \quad C_c \in \mathbf{R}^{m_p \times n_c}, \quad D_c \in \mathbf{R}^{m_p \times r_p}.$$

Using the augmented state $x = \begin{bmatrix} x_p \\ x_c \end{bmatrix}$, we obtain for $r = 0$,

$$\dot{x} = A_F x \tag{7.43}$$

where

$$\begin{aligned} A_F &= \begin{bmatrix} A_p + B_p D_c C_p & B_p C_c \\ B_c C_p & A_c \end{bmatrix} \\ &= \begin{bmatrix} A_p & 0 \\ 0 & 0 \end{bmatrix} + \begin{bmatrix} B_p & 0 \\ 0 & I \end{bmatrix} \begin{bmatrix} D_c & C_c \\ B_c & A_c \end{bmatrix} \begin{bmatrix} C_p & 0 \\ 0 & I \end{bmatrix}. \end{aligned} \tag{7.44}$$

Thus,

$$A_F = A + BFC \tag{7.45}$$

where

$$A = \begin{bmatrix} A_p & 0 \\ 0 & 0 \end{bmatrix}, \quad B = \begin{bmatrix} B_p & 0 \\ 0 & I_{n_c} \end{bmatrix}, \quad C = \begin{bmatrix} C_p & 0 \\ 0 & I_{n_c} \end{bmatrix},$$

$$F = \begin{bmatrix} D_c & C_c \\ B_c & A_c \end{bmatrix},$$

$$A \in \mathbf{R}^{(n_p + n_c) \times (n_p + n_c)}, \quad B \in \mathbf{R}^{(n_c + n_p) \times (n_c + m_p)}$$

$$C \in \mathbf{R}^{(n_c + r_p) \times (n_p + n_c)}, \quad F \in \mathbf{R}^{(n_c + m_p) \times (n_c + r_p)}. \tag{7.46}$$

The closed loop characteristic polynomial is

$$\Delta(s) = | \, sI - (A + BFC) \, | . \tag{7.47}$$

Our objective is to design a compensator so as to meet some specified performances. We do not intend to present a complete feedback design. Rather, we wish to concentrate on the application of root clustering to feedback design. To this end, we suppose that C(s) has a proper structure for tracking. For instance, for step tracking, we need a free integrator in $C(s)$, $C(s) = \frac{1}{s}\, C_1(s)$. All that is left is the asymptotic stability of the closed loop. We may state the following objective.

Design Objective 7.1 Find $\{a_i, b_i\}$, or F, such that $\Delta(s)$ is asymptotically stable.

One way to solve this problem is to use pole placement. Toward this end, consider once more Figure 7.1 . For n-th order plant, deg $[d(s)] = n$, and for asymptotic stability (without tracking), the compensator's order is n-1. Indeed, let

$$a(s) = \sum_{i=0}^{n-1} a_i s^i \qquad b(s) = \sum_{i=0}^{n-1} b_i s^i \tag{7.48}$$

and let the required closed loop characteristic polynomial be

$$\Delta(s) = \sum_{i=0}^{2n-1} \Delta_i s^i . \tag{7.49}$$

Then (7.23) implies

$$[a_{n-1} \; a_{n-2}, \ldots a_0 \quad b_{n-1} \; b_{n-2}, \ldots, b_0] \; S(n,d) = [\Delta_{2n-1} \; \Delta_{2n-2}, \ldots, \Delta_1 \; \Delta_0] \tag{7.50}$$

where S(n,d) is the Sylvester matrix of n(s) and d(s) given by (1.20). We know from Theorem 5.2 that S(n,d) is nonsingular so that (7.35) has a unique solution for $\{a_i\}$, $\{b_i\}$ if and only if n(s) and d(s) are coprime (G(s) is minimal). Because of its simplicity this synthesis approach is attractive.

However, as shortly will be demonstrated, this method suffers two main drawbacks. First, in many cases a ***compensator's order*** of n-1 is not acceptable. Second, it may lead to the ***inverse response*** which affects step response like time delay. This response is characterized by right hand side zeros in the compensator.

Example 7.3

Let $G(s) = \dfrac{1}{1+s/14.3} \cdot \dfrac{1}{1+2 \cdot 0.2(s/3)+(s/3)^2}$. We seek a compensator C(s) so as to track a step command. To this end, we require C(s) to contain a pure integrator. This implies that $\Delta(s)$ is of order 2n (instead of 2n-1 as in the above discussion). We choose the closed loop poles

$$-0.2, \quad -0.72 \pm j2.8, \quad -9.73 \pm j\,19.5, \quad -13.9\,,$$

and obtain the unique n-th order (instead of n-1) compensator

$$C(s) = \frac{0.3s^3 + 20s^2 - 8.8s + 93.2}{s(s^2 + 19.5s + 483)}\,.$$

Although the closed loop system is asymptotically stable, the step response is not acceptable. The reason is the ***inverse response***. That is, first the output propagates in an opposite direction to the step command, and only after some time it tends to its final direction. This behaviour has similarities with time delay, and should be avoided whenever possible. To see the reason for the inverse response, note that C(s) obtained in Example 7.3 is a nonminimum phase. That is, C(s) has zeros in the right half plane (a(s) is not Hurwitz). Clearly, since using pole placement, C(s) of order n is unique, the inverse response may occur. On the other hand, if we choose (among many possibilities)

$$C(s) = \frac{k\,(s^2 + 1.2s + 9)}{s(s + 14.3)}$$

with $0 < k < 8.7$, the closed loop is asymptotically stable.

It is clear that the undesirable response is caused by the requirement for ***specific locations*** of the closed loop poles. In contrast, for ***good behavior*** we only need ***pole/zero clustering*** in some regions in the complex plane. The freedom we now have, enables us to add one more requirement, namely the reduction of the compensator's order. We now modify Design Objective 7.1 as follows.

Design Objective 7.2 For the basic feedback configuration in Figure 1, find $\{a_i, b_i\}$ such that

(i) $\sigma(\Delta) \subset \aleph_1$,

(ii) $\sigma(a) \subset \aleph_2$,

(iii) b(s) has the lowest degree possible.

To solve this design problem, we use the following algorithm.

Algorithm 7.2 (Compensator's Construction, Figure 7.1)

Data: (i) Minimal transfer function G(s), or polynomials n(s) and d(s).

(ii) Root clustering region $\aleph_1$ for relative stability.

(iii) Region $\aleph_2$, usually the left half plane.

Step 0: i=0

Step 1: Set deg(b) = i. Note for i=0, $C(s) = a_0/b_0$ is proportional control.

Step 2: Calculate $\Delta(s)$ according to (7.35)

Step 3: Apply Algorithm 7.1 to find $\{a_i, b_i)$ such that $\sigma(\Delta) \subset \aleph_1$ and $\sigma(a) \subset \aleph_2$.
Note that c(p) is the product of the respective critical constraints.

Step 4: If solution exists — stop. Otherwise, set i = i+1 and go to step 1.

7.5 COMPENSATOR DESIGN — MATRIX FORM

In the above section we discussed compensator design in polynomial form, based on the critical polynomial and its derivatives, see (7.32). This design may achieve (if possible) closed loop relative stability, stable compensators and compensators with left half plane zeros. Moreover, we can handle feedback configurations where both the plant and the compensator transfer functions are proper, and we can fix the compensator structure. However, since the compensator's parameters enter the critical polynomial implicitly, the set of equations (7.32) is not structured. To obtain a structured set of equations, we develop now a matrix version. Toward this end, consider the feedback system (7.43). Note that here the plant must be strictly proper. The objective is to select a compensator F such that $\sigma(A_F) \subset \aleph$. Before discussing a general region $\aleph$, we start with the left half plane for asymptotic stability. A close look at Example 2.1 reveals that we have solved the minimization problem

$$\begin{aligned} &\min_F \ \mathrm{tr}(P) \\ &\text{s. t.} \quad PA_F + A'_F P + Q = 0 . \end{aligned} \tag{7.51}$$

In order to prevent the possibility of minimum at infinity, we add a quadratic form to tr(P),

$$\begin{aligned} &\min_F \ \mathrm{tr}(P + F'MF) \\ &\text{s.t.} \quad PA_F + A'_F P + Q = 0 . \end{aligned} \tag{7.52}$$

More general, suppose region $\aleph$ is M-transformable, and we wish to select a compensator F such that the spectrum of A_F lies in $\aleph$. In this general case, we construct the minimization problem

$$\begin{aligned} &\min_F J = \mathrm{tr}(RP + F'MF) \\ &\text{s.t.} \quad s = \sum_{i,j} \phi_{ij} A_F^i P A_F^{'j} + Q = 0 , \end{aligned} \tag{7.53}$$

where $R \in \mathbf{R}^{(n_p+n_c)\times(n_p+n_c)}$, $M \in \mathbf{R}^{(n_c+m_p)\times(n_c+m_p)}$, $Q \in \mathbf{R}^{(n_p+n_c)\times(n_p+n_c)}$ are given p.d. symmetric matrices. Define the Lagrangian

$$L = \mathrm{tr}\left[RP + F'MF + \Lambda\left(\sum_{i,j} \phi_{ij} A_F^i P A_F^{'j} + Q\right)\right], \tag{7.54}$$

where $\Lambda \in \mathbf{R}^{(n_c+m_p)\times(n_c+m_p)}$ is the Lagrange multiplier matrix. Then, using the Appendix, the necessary conditions for minimum are

$$\text{(i)} \quad \frac{\partial L}{\partial \Lambda} = \sum_{i,j} \phi_{ij} A_F^i P A_F^{'j} + Q = 0 \quad , Q > 0$$

$$\text{(ii)} \quad \frac{\partial L}{\partial P} = \sum_{i,j} \phi_{ij} A_F^{'i} \Lambda A_F^j + R = 0 \quad , R > 0 \tag{7.55}$$

$$\text{(iii)} \quad \frac{\partial L}{\partial F} = \sum_{i,j} \phi_{ij}\left(\sum_{k=0}^{i-1} B' A_F^{'k} \Lambda A_F^j P A_F^{'i-1-k} C' + \sum_{k=0}^{j-1} B' A_F^{'k} \Lambda A_F^i P A_F^{'j-1-k} C'\right) + 2MF = 0, \quad M > 0.$$

Theorem 7.3 Consider the feedback system (7.43) with a given strictly proper plant, and an M-Transformable region $\aleph$. A fixed order compensator F exists such that the closed loop satisfies $\sigma(A_F) \subset \aleph$, if and only if (7.55) admits a solution $\{F, \Lambda, P\}$ satisfying $\Lambda > 0$, $P > 0$.

Proof Sufficiency. If (7.55) admits a solution satisfying $\Lambda > 0$, $P > 0$, it follows that (7.55i) is satisfied with $P > 0$. Thus by Theorem 6.5, $\sigma(A_F) \subset \aleph$.

Necessity. Let $\tilde{\aleph}$ be the image of the given region $\aleph$ in the parameter space spanned by F. Suppose $\tilde{\aleph}$ is nonempty. That is, there are F's satisfying $\sigma(A_F) \subset \aleph$. Then by Theorem 6.5, for each such F, the equality constraint in (7.53) satisfies $P > 0$, so that the cost in (7.53) is a finite positive number. On the other hand, as $F \rightarrow \partial\tilde{\aleph}$, that is, as we approach a finite boundary in the parameter space, $\mathrm{tr}(P) \rightarrow \infty$, so that the cost approaches infinity. This can be verified by (6.7) for a simple matrix A_F.

If $\tilde{\aleph}$ has boundary at infinity, the second term in the cost in (7. 53) approaches infinity. We conclude that $J = \infty$ on $\partial\tilde{\aleph}$. In addition, $J(\cdot)$ is smooth, thus a minimum exists. Since $\aleph$ is M-transformable and so $\phi(\lambda_i, \bar{\lambda}_j) \neq 0$ in $\aleph$, it follows that in $\tilde{\aleph}$, , $s = 0$ is solvable for P uniquely; see (6.11). Thus, n^2 variables can be expressed by the rest, and necessary conditions for minimum are given by $\mathrm{grad}L = 0$. In other words, (7. 55) must hold somewhere in the nonempty set $\tilde{\aleph}$. In the case that A_F is not simple, we may apply a limit argument.. This completes the proof.

Example 7.4 Let $\aleph$ be the left half plane. In this case (7.53) becomes

$$\min_F J = \mathrm{tr}(RP + F'MF)$$
$$\text{s.t.} \quad s = PA_F + A'_F P + Q = 0 \tag{7.56}$$

and (7.55) becomes

$$\text{(i)} \quad \frac{1}{2}\frac{\partial L}{\partial F} = MF + B'P\Lambda C' = 0\,,$$
$$\text{(ii)} \quad \frac{\partial L}{\partial \Lambda} = PA_F + A'_F P + Q = 0\,, \tag{7.57}$$
$$\text{(iii)} \quad \frac{\partial L}{\partial P} = \Lambda A'_F + A_F \Lambda + R = 0\,.$$

From (7.57i), it follows that

$$F = -M^{-1}B'P\Lambda C' \tag{7.58}$$

where, from (ii), (iii), we have

$$\text{(i)} \quad PA + A'P - PBM^{-1}B'P\Lambda C'C - C'C\Lambda PBM^{-1}B'P + Q = 0, \tag{7.59}$$
$$\text{(ii)} \quad \Lambda A' + A\Lambda - \Lambda C'C\Lambda PBM^{-1}B' - BM^{-1}B'P\Lambda C'C\Lambda + R = 0\,.$$

Thus, we first solve (7.59) for $P > 0$ and $\Lambda > 0$, and then substitute in (7.58) to obtain F.

Example 7.5 Let $\aleph$ be the unit circle. In this case (7.53) becomes

$$\min_F J = \mathrm{tr}(RP + F'MF)$$
$$\text{s.t.} \quad s = A_F P A'_F - P + Q = 0 \tag{7.60}$$

and (7.55) becomes

$$\text{(i)} \quad \frac{1}{2}\frac{\partial L}{\partial F} = MF + B'\Lambda A_F PC' = 0\,,$$
$$\text{(ii)} \quad \frac{\partial L}{\partial \Lambda} = A_F P A'_F - P + Q = 0\,, \tag{7.61}$$
$$\text{(iii)} \quad \frac{\partial L}{\partial P} = A'_F \Lambda A_F - \Lambda + R = 0\,.$$

From (7.61 i)

$$MF + B'\Lambda(A + BFC)PC' = 0$$

or

$$MF + \Lambda_B F P_c + B'\Lambda A P C' = 0 , \tag{7.62}$$

where

$$\Lambda_B = B'\Lambda B, \qquad P_c = C'PC .$$

Since M is an arbitrary symmetric matrix, ***choose***

$$M = B'\Lambda B. \tag{7.63}$$

Thus, from (7.62)

$$F = - \Lambda_B^{-1} B'\Lambda A P C' (I + P_c)^{-1} \tag{7.64}$$

Substituting F in (7.61 ii) and (7.61 iii), we obtain two matrix equations in the two unknown matrices P and Λ.

An alternative statement to the minimization problem (7.53) is the following.

$$\underset{F}{\text{Min}} \;\; J = \text{tr} \;\; [(R + C'F'VFC)P]$$
$$\text{s.t.} \quad s = \sum_{i,j} \phi_{ij} A_F^i P A_F^{'j} + Q + BFWF'B' = 0 , \tag{7.65}$$

where R, V, Q, W are given p.d. symmetric matrices. Define the Lagrangian

$$L = \text{tr}\left[(R + C'F'VFC)P + \Lambda \left(\sum_{i,j} \phi_{ij} A_F^i P A_F^{'j} + Q + BFWF'B' \right) \right] . \tag{7.66}$$

The corresponding necessary conditions for minimum, are

$$\text{(i)} \quad \frac{\partial L}{\partial \Lambda} = \sum_{i,j} \phi_{ij} A_F^i P A_F^{'j} + Q + BFWF'B' = 0 ,$$

$$\text{(ii)} \quad \frac{\partial L}{\partial P} = \sum_{i,j} \phi_{ij} A_F^{'i} \Lambda A_F^j + R + C'F'VFC = 0 , \tag{7.67}$$

$$\text{(iii)} \quad \frac{\partial L}{\partial F} = \sum_{i,j} \phi_{ij} \left(\sum_{k=0}^{i-1} B'A_F^{'k} \Lambda A_F^j P A_F^{i-1-k} C' + \sum_{k=0}^{j-1} B'A_F^{'k} \Lambda A_F^i (A_F^{'})^{j-1-k} C' \right)$$

$$+ 2VFCPC' + 2B'\Lambda BFW = 0 .$$

Theorem 7.4 Consider the feedback system (7.43) with a given strictly proper plant, and an M-transformable region $\aleph$. A fixed order compensator F exists such that the closed loop satisfies $\sigma(A_F)\subset\aleph$, if and only if (7.67) admits a solution $\{F,\Lambda,P\}$ satisfying $\Lambda>0,\ P>0$.

Thus far, we have used minimization for compensator design. Computationally, we have to solve a set of polynomial equations (7.55), or (7.67). Without discussing methods of generating solutions, it is evident that we expect the above set to have a finite number of (complex) solutions. However, we do not have any way to guarantee this expectation. In fact, in general, such a set may have an infinite number of solutions. In order to assure a finite number of solutions, we combine the results of the present section with those of Section 7.4. We modify (7.53) as follows.

$$\begin{aligned}
&\min_F\ J=\mathrm{tr}[(F-F_o)'M(F-F_o)]+(t-\tau)^2\\
&\text{s.t.}\ \ s_1=\mathrm{tr}(RP)-t^2=0\\
&\qquad s_2=\sum_{i,j}\phi_{ij}A_F^iPA_F^{'j}+Q=0\,.
\end{aligned}\tag{7.68}$$

The Lagrangian becomes

$$\begin{aligned}
L=\mathrm{tr}[(F-F_o)'M(F-F_o)]+(t-\tau)^2+\lambda(\mathrm{tr}(RP)-t^2)\\
+\mathrm{tr}\left[\Lambda\Big(\sum_{i,j}\phi_{ij}A_F^iPA_F^{'j}+Q\Big)\right]
\end{aligned}\tag{7.69}$$

and the necessary conditions for minimum are

$$\begin{aligned}
&\text{(i)}\quad \frac{\partial L}{\partial\Lambda}=\sum_{i,j}\phi_{ij}A_F^iPA_F^{'j}+Q=0\,,\\
&\text{(ii)}\quad \frac{\partial L}{\partial\lambda}=\mathrm{tr}(RP)-t^2=0\,,\\
&\text{(iii)}\quad \frac{\partial L}{\partial P}=\sum_{i,j}\phi_{ij}A_F^{'i}\Lambda A_F^j+\frac{t-\tau}{t}R=0\,,\\
&\text{(iv)}\quad \frac{\partial L}{\partial F}=\sum_{i,j}\phi_{ij}\left(\sum_{k=0}^{i-1}B'A_F^{'k}\Lambda A_F^jPA_F^{'i-1-k}C'+\sum_{k=0}^{j-1}B'A_F^{'k}\Lambda A_F^iPA_F^{'j-1-k}C'\right)\\
&\qquad\qquad +2M(F-F_0)=0\,.
\end{aligned}\tag{7.70}$$

Note that for $\tau=0$ and $F_0=0$, system (7.70) reduces to (7.55). Note also that (7.70iii) contains $\lambda=\frac{t-\tau}{t}$. This is evident from $\frac{\partial L}{\partial t}=0$.

Theorem 7.5 Consider the feedback system (7.43) with a given strictly proper plant, and an M-transformable region $\aleph$. A fixed order compensator F exists such that the closed loop satisfies $\sigma(A_F)\subset\aleph$, if and only if (7.70) admits a solution $\{F,\Lambda,P,t\}$ satisfying $\Lambda>0,\ P>0$. For almost all $\{F_0,\tau\}$, system (7.70) has a finite number of solutions.

7.6 ROBUST ANALYSIS

In previous sections we have constructed compensators under the assumption that the plant is completely known. In real life, however, we know the parameter values with some uncertainty. In this section we assume that a given nominal system is stable with respect to a region $\aleph$ in the complex plane. We are also given a compact set Ω in the parameter space $\mathbf{R}^m$. Our objective is to verify the stability of the system for all parameter values inside Ω. To this end, consider the uncertain characteristic polynomial

$$\Delta(\lambda; p) = \sum_{i=0}^{n} a_i(p)\lambda^i \quad ; \quad a_i(p) \in \mathbf{R}[p_1, p_2, \dots p_m] \tag{7.71}$$

where $p \in \mathbf{R}^m$ is the uncertain parameter vector. We assume that p is restricted according to

$$p \in \Omega \subset \mathbf{R}^m; \quad \Omega = \{ p \in \mathbf{R}^m : p' Mp - d^2 \leq 0 \} . \tag{7.72}$$

For simplicity, we assume that M is diagonal with $m_{ii} > 0$. Thus, Ω can be written as

$$\Omega = \left\{ p \in \mathbf{R}^m : \sum_{i=1}^{m} m_i p_i^2 \leq d^2 \right\} . \tag{7.72)'$$

We further assume that the nominal system is stable with respect to $\aleph$; that is,

$$\sigma(\Delta(\lambda; 0)) \subset \aleph . \tag{7.73}$$

Using ideas from Section 7.2, we have the following result.

Theorem 7.6 Consider the uncertain polynomial $\Delta(\lambda;p)$ given by (7.71), and suppose the nominal polynomial $\Delta(\lambda;0)$ has all its roots inside $\aleph$, a given H-transformable region (Definition 7.2). Then, $\sigma(\Delta(\lambda;p)) \subset \aleph \;\; \forall\, p \in \Omega$, if and only if $c(p) > 0 \;\; \forall\, p \in \Omega$.

Proof A direct consequence of (7.14) and the continuity of the roots with respect to the parameters.

To test the positivity $c(p)>0 \;\; \forall\, p \in \Omega$, we may calculate the minimum value of $c(p)$ in Ω. This minimum must be positive. Note, however, that a direct calculation of the minimum may lead to a set of equations with an infinite number of solutions. This prevents us from isolating the global minimum. To avoid this difficulty, we use a different approach. Consider the minimization problem

$$\text{Min} \sum_{i=1}^{m} (p_i - \gamma_i)^2 + \sum_{i=1}^{2} (t_i - \tau_i)^2$$

$$\text{s.t.} \quad s_1 = c(p) + t_1^2 = 0 \tag{7.74}$$

$$s_2 = \sum_{i=1}^{m} m_i p_i^2 - d^2 + t_2^2 = 0$$

where $\{\gamma_i;\ i = 1,2,\dots,m\}$ and $\{\tau_i;\ i = 1,2\}$ are constants, chosen randomly.
Clearly, if $c(p) > 0\ \forall\ p \in \Omega$, then there exists no real solution to $s_1 = 0,\ s_2 = 0$. Conversely, if there exists no solution to $s_1 = 0,\ s_2 = 0$, then at points where $s_2 = 0$ ($p \in \Omega$), s_1 must be nonzero, or $c(p) \neq -t_1^2$; thus $c(p) > 0\ \forall\ p \in \Omega$. We conclude with the following theorem.

Theorem 7.7 $c(p) > 0\ \forall\ p \in \Omega$, if and only if the minimization (7.74) has no real solution.

Proof The minimization (7.74) is simply the minimum distance from a point (p,t_1,t_2) to the surface $\{s_1 = 0,\ s_2 = 0\}$. If (7.74) has no real solution so does $\{s_1 = 0,\ s_2 = 0\}$, and vice versa. Using the discussion above, the theorem follows.

To solve (7.74), form the Lagrangian

$$L = \sum_{i=1}^{m} (p_i - \gamma_i)^2 + \sum_{i=1}^{2} (t_i - \tau_i)^2 + \lambda_1(c(p) + t_1^2) + \lambda_2\left(\sum_{i=1}^{m} m_i p_i^2 - d^2 + t_2^2\right).$$

Then, necessary conditions for minimum, are

$$\begin{aligned}
&\text{(i)} \quad \frac{\partial L}{\partial p_i} = 2(p_i - \gamma_i) + \lambda_1 \frac{\partial c(p)}{\partial p_i} + 2\lambda_2 m_i p_i = 0 \quad ; \quad i = 1, 2, \dots, m \\
&\text{(ii)} \quad \frac{\partial L}{\partial \lambda_1} = c(p) + t_1^2 = 0 \\
&\text{(iii)} \quad \frac{\partial L}{\partial \lambda_2} = \sum_{i=1}^{m} m_i p_i^2 - d^2 + t_2^2 = 0 \\
&\text{(iv)} \quad \frac{\partial L}{\partial t_i} = 2(t_i - \tau_i) + 2\lambda_i t_i \quad ; \quad i = 1, 2.
\end{aligned} \qquad (7.75)$$

Note that (7.75) is a set of m+4 polynomial equations in m+4 variables. This can be reduced to m + 2 equations by solving (iv) for λ_1 and λ_2.

Combining Theorems 7.6, 7.7 and the set of equations (7.75), we state the following theorem.

Theorem 7.8 Consider the uncertain polynomial $\Delta(\lambda;p)$ given by (7.71), and suppose the nominal polynomal $\Delta(\lambda;0)$ has all its roots inside $\aleph$, a given H-transformable region (Definiton 7.2). Then $\sigma(\Delta(\lambda;p)) \subset \aleph\ \forall\ p \in \Omega$, if and only if the set of equations (7.75) has no real solution. Moreover, this set has a finite number of solutions (real and complex) for almost all $\gamma_1,\gamma_2,\dots,\gamma_m,\ \tau_1,\ \tau_2$.

Next, we discuss a matrix version of the robust analysis. Consider the feedback system (7.43) - (7.45), where F is a fixed compensator and A,B,C, are the plant parameters. We assume that A,B,C, contain uncertainty according to

$$A = A_o + \Delta A, \quad B = B_o + \Delta B, \quad C = C_o + \Delta C \qquad (7.76)$$

where A_o, B_o, C_o, are the nominal values. In particular,

$$\Delta A = \begin{bmatrix} \Delta A_p & 0 \\ 0 & 0 \end{bmatrix} = \begin{bmatrix} I \\ 0 \end{bmatrix} \Delta A_p [\, I \quad 0 \,]$$

$$\Delta B = \begin{bmatrix} \Delta B_p & 0 \\ 0 & 0 \end{bmatrix} = \begin{bmatrix} I \\ 0 \end{bmatrix} \Delta B_p [\, I \quad 0 \,]$$

$$\Delta C = \begin{bmatrix} \Delta C_p & 0 \\ 0 & 0 \end{bmatrix} = \begin{bmatrix} I \\ 0 \end{bmatrix} \Delta C_p [\, I \quad 0 \,] . \tag{7.77}$$

We restrict the uncertainty as follows.

$$\mathrm{tr}[(\Delta A_p)' M_1 (\Delta A_p)] + \mathrm{tr}[(\Delta B_p)' M_2 (\Delta B_p)] + \mathrm{tr}[(\Delta C_p)' M_2 (\Delta C_p)] - d^2 \leq 0 \tag{7.78}$$

where M_1, M_2, M_3 are all diagonal positive definite. Recalling (6.7), on the boundary $\partial\aleph$, for some λ_i, $\phi(\lambda_i, \bar{\lambda}_i)$ vanishes, and thus P_{ii} tends to infinity. Thus, $P_{ii}^{-1} = 0$ on $\partial\aleph$ for some i. We conclude that $(\mathrm{tr}RP)^{-1}$ behaves like the critical constraint $c(p)$ – it vanishes on $\partial\tilde{\aleph}$ and is positive inside $\tilde{\aleph}$. Thus, the matrix version of (7.74) is the following minimization problem.

$$\text{Min} \sum_{i=1}^{3} \mathrm{tr}(\Delta_i - \Gamma_i)^1 (\Delta_i - \Gamma_i) + \sum_{i=1}^{2} (t_i - \tau_i)^2$$

$$\text{s.t.} \quad s_1 = (\mathrm{tr}RP)^{-1} + t_1^2 = 0$$

$$s_2 = \sum_{i,j} \phi_{ij} A_F^i P A_F^{'j} + Q = 0 \tag{7.79}$$

$$s_3 = \sum_{i=1}^{3} \mathrm{tr}(\Delta_i' M_i \Delta_i) - d^2 + t_2^2 = 0$$

where $\Delta_1 = \Delta A_p$, $\Delta_2 = \Delta B_p$, and $\Delta_3 = \Delta C_p$.

7.7 INTERVAL POLYNOMIALS AND PLANTS

In the previous section we have discussed a general approach to robustness. However, numerical procedures are not easy to implement. In the special case where the parameter space reduces to the real line, that is, the system contains a single parameter, the computation is straightforward as we now show. Consider the polynomial

$$\Delta(\lambda; \gamma) = \sum_{i=0}^{n} a_i(\gamma) \lambda^i \quad , \tag{7.80}$$

where $a_i(\gamma)$ are given polynomials in the single parameter γ. Note that we do not restrict $a_i(\cdot)$ in any way. The parameter γ can be thought of as a controller parameter to be selected, or as an uncertain parameter of the plant. We wish to find the intervals of γ in which the system is stable with respect to $\aleph$. For this case, we do not use the general Algorithm 7.1. Instead we use the following steps:

Algorithm 7.3 (a single parameter system)

1. Generate the critical polynomial $c(\gamma)$.
2. Find numerically all the real roots of $c(\gamma)$.

   These roots decompose the real axis into open intervals $\{\Gamma_i ;\ i \in \mathbf{J}\}$, where $\mathbf{J}$ is a proper index set.

3. Pick an arbitrary point in Γ_i, say γ_{io}, and test $\sigma(\Delta(\lambda;\gamma_{io}))\subset \aleph$ using root clustering theory or by numerical means.
4. If $\gamma_i \in \tilde{\aleph}$, then $\Gamma_i \subset \tilde{\aleph}$. If $\gamma_i \notin \tilde{\aleph}$, then Γ_i is excluded from $\tilde{\aleph}$.

Example 7.6 Consider the polynomial $\Delta(\lambda;\gamma) = \gamma\lambda^4 + (3+\gamma)\lambda^3 + (4+\gamma)\lambda^2 + 4\lambda + 3.5$ with $\gamma \in [2, 16]$. We wish to check the stability of $\Delta(\lambda;\gamma)$ with respect to the left half plane. Note that for $\gamma = 2$ and for $\gamma = 16$, $\Delta(\lambda;\gamma)$ is stable. However, as we now show, it is not sufficient. Using Algorithm 7.3, we find the critical constraint,

$$c(\gamma) = \left|H_4(\gamma)\right| = \begin{vmatrix} 3+\gamma & 4 & 0 & 0 \\ \gamma & 4+\gamma & 3.5 & 0 \\ 0 & 3+\gamma & 4 & 0 \\ 0 & \gamma & 4+\gamma & 3.5 \end{vmatrix} = 3.5(0.5\gamma^2 - 9\gamma + 16.5)\,.$$

The zeros of $c(\gamma)$, are: 2.07 and 15.93. However, for $\gamma = 10$ we find $c(10) < 0$. Thus, for $\gamma \in [2.07, 15.93]$, $\Delta(\lambda;\gamma)$ is unstable.

We now move to another extreme case. We consider the polynomial

$$\Delta(\lambda; a) = \sum_{i=0}^{n} a_i\lambda^i \quad ; \quad a_n > 0 \tag{7.81}$$

where $a = [a_o \ldots a_n]'$ is the coefficients vector, with

$$a_i^- \le a_i \le a_i^+ \quad ; \quad i = 0, 1, \ldots, n\,. \tag{7.81'}$$

We call such a polynomial an ***interval polynomial*.**

Theorem 7.9 The entire family of polynomials in (7.81) has all its zeros in the open left half plane, if and only if each of the following four polynomials has all its zeros in the open left half plane:

$$K_1(\lambda) = a_o^- + a_1^-\lambda + a_2^+\lambda^2 + a_3^+\lambda^3 + a_4^-\lambda^4 + a_5^-\lambda^5 + a_6^+\lambda^6 + \dots$$

$$K_2(\lambda) = a_o^+ + a_1^+\lambda + a_2^-\lambda^2 + a_3^-\lambda^3 + a_4^+\lambda^4 + a_5^+\lambda^5 + a_6^-\lambda^6 + \dots$$

$$K_3(\lambda) = a_o^+ + a_1^-\lambda + a_2^-\lambda^2 + a_3^+\lambda^3 + a_4^+\lambda^4 + a_5^-\lambda^5 + a_6^-\lambda^6 + \dots$$

$$K_4(\lambda) = a_o^- + a_1^+\lambda + a_2^+\lambda^2 + a_3^-\lambda^3 + a_4^-\lambda^4 + a_5^+\lambda^5 + a_6^+\lambda^6 + \dots \quad . \tag{7.82}$$

Proof First, we introduce the real and imaginary parts of the polynomials $K_i(\lambda)$:

$$\begin{aligned} R^-(\omega) &= \text{Re } K_1(j\omega) \\ R^+(\omega) &= \text{Re } K_2(j\omega) \\ I^-(\omega) &= \text{Im } K_3(j\omega) \\ I^+(\omega) &= \text{Im } K_4(j\omega) . \end{aligned} \tag{7.83}$$

As an illustration, for $n = 5$, we obtain

$$R^-(\omega) = a_4^-\omega^4 - a_2^+\omega^2 + a_o^-$$

$$R^+(\omega) = a_4^+\omega^4 - a_2^-\omega^2 + a_o^+$$

$$I^-(\omega) = a_5^-\omega^5 - a_3^+\omega^3 + a_1^-\omega$$

$$I^+(\omega) = a_5^+\omega^5 - a_3^-\omega^3 + a_1^+\omega \quad . \tag{7.83'}$$

Now, observe that in $R^-(\omega)$, the coefficients a_4 and a_o which appear with a + sign take the minimal values a_4^- and a_o^-, while a_2 which appears with a - sign takes the maximal value a_2^+. Continuing this argument, we obtain

For $\omega \geq 0$,

$$\begin{aligned} R^-(\omega) &\leq \text{Re } \Delta(j\omega;a) \leq R^+(\omega) \\ I^-(\omega) &\leq \text{Im } \Delta(j\omega;a) \leq I^+(\omega) . \end{aligned} \tag{7.84}$$

For $\omega < 0$,

$$\begin{aligned} R^-(\omega) &\leq \text{Re } \Delta(j\omega;a) \leq R^+(\omega) \\ I^+(\omega) &\leq \text{Im } \Delta(j\omega;a) \leq I^-(\omega) . \end{aligned} \tag{7.84'}$$

Thus, (7.84) describes a box in which $\Delta(j\omega;a)$ lies for all coefficient values satisfying the box constraint (7.81)'. Next, according to the separation property of a Hurwitz (stable) polynomial, the polar plot of $\Delta(j\omega)$ cuts the real and the imaginary axes alternately a total of n times. Thus, if the polynomials $K_i(\lambda)$, i = 1,...,4, are all stable, it follows that the four box vertices in (7.84) satisfy the above separation property. By continuity of the polar plot with respect to parameters it now follows that the entire box has the same property. This proves the sufficiency part of the theorem. The necessity part is trivial. This completes the proof.

Example 7.7 Let the box constraint (7.81)' be given by

$$a_i^- = a_{oi} - \gamma_i \varepsilon \quad , \quad a_i^+ = a_{oi} + \delta_i \varepsilon \quad , \quad i = 0, 1, 2, \dots, n \tag{7.85}$$

with $\varepsilon > 0$, $\gamma_i > 0$, $\delta_i > 0$. We assume that the nominal polynomial $\Delta(\lambda;a_o)$ is stable, and we seek the largest value of ε, say ε_{max}, for which the uncertain system is stable. Using the idea of the critical constraint (Section 7.2), and in particular (7.16), we define the four critical constraints $c_i(\varepsilon)$, i = 1, 2, 3, 4. Denote by ε_i^{min} the smallest positive real root of $c_i(\varepsilon)$. Then

$$\varepsilon_{max} = \min\left\{ \varepsilon_i^{min}, \ i = 1, 2, 3, 4 \right\} . \tag{7.86}$$

Theorem 7.9 is important since the test for robustness reduces to simple stability tests of four polynomials with constant coefficients. This result, however, suffers two drawbacks. First, it is limited to the left half plane. Second, the structure of interval polynomials implies independent coefficient perturbations. Said another way, in the uncertain polynomial (7.71), no p_i enters into more than one coefficient a_i. This limitation prevents us from using the above theorem in feedback analysis.

Example 7.8 Consider the basic feedback system as in Figure 7.1, with c(s) = 3/(s+1) and $G(s) = s/(1-s+\alpha s^2 + s^3)$. Assume that the parameter α is uncertain according to $\alpha \in [3.4, 5]$. It is readily seen that the nominal closed loop with $\alpha = 4$ is stable with respect to the left half plane. The closed loop characteristic polynomial is

$$\Delta(s;\alpha) = s^4 + (\alpha + 1)s^3 + (\alpha - 1)s^2 + 3s + 1 .$$

We see that the coefficients $a_3 = \alpha + 1$ and $a_2 = \alpha - 1$ are not independent. In fact, $a_3 = a_2 + 2$.. Moreover, if we take a_3 and a_2 as independent, with $a_3 \in [4.4, 6]$ and $a_2 \in [2.4, 4]$, then using (7.82), we have $K_3(s) = s^4 + 6s^3 + 2.4s^2 + 3s + 1$. Since $K_3(s)$ is unstable, we may conclude that $\Delta(s;\alpha)$ is unstable for $\alpha \in [3.4,5]$. However, $c(\alpha) = |H_4(\alpha)| = 2\alpha^2 - 2\alpha - 13$. It is readily found that $\Delta(s;\alpha)$ is stable for all $\alpha > 3.1$; thus, for this simple feedback system Theorem 7.9 cannot be used.
Our first generalization is concerned with the feedback structure. Consider the basic feedback configuration as in Figure 7.1. If G(s) = n(s)/d(s) and C(s) = a(s)/b(s), the closed loop characteristic polynomial becomes $\Delta(s) = a(s)n(s) + b(s)d(s)$ where a(s) and b(s) are fixed polynomials. We assume that the plant is uncertain such that n(s) and d(s) are interval polynomials. For the sake of single input multi output or multi input single output, consider the characteristic polynomial

$$\Delta(s) = \sum_{i=1}^{r} a_i(s) n_i(s) \tag{7.87}$$

where the polynomials $a_i(s)$ are fixed, whereas the polynomials $n_i(s)$ are interval polynomials.

Plants that have such properties are called **interval plants**. Let $\underline{N} = (n_1(s), n_2(s) \ldots, n_r(s))$ and $\underline{A} = (a_1(s), a_2(s), \ldots, a_r(s))$ be two r–tuples of real polynomials. Then, we can write (7.87) as $\underline{\Delta} = \underline{A}\,\underline{N}$.

With each interval polynomial $n_i(s)$ we associate the four corner polynomials defined in (7.82), $K_{i1}(s)$, $K_{i2}(s)$, $K_{i3}(s)$, and $K_{i4}(s)$. We now define $m4^m$ line segments as follows. For any fixed integer k between 1 and m, set

$$n_i(s) = K_{ij}(s), \quad \text{for } i \neq k \text{ and for some } j = 1, 2, 3, 4$$

and for k, suppose that $n_i(s)$ varies in one of the four segments

$$[K_{k1}(s), K_{k2}(s)],\ [K_{k1}(s), K_{k3}(s)],\ [K_{k2}(s), K_{k4}(s)],\ [K_{k3}(s), K_{k4}(s)].$$

By segment $[K_{k1}(s), K_{k2}(s)]$ we mean all convex combinations of the form $(1-\mu)K_{k1}(s)+\mu K_{k2}(s)$ with $\mu \in [0, 1]$. There are indeed $m4^m$ such segments. Any one of these segments is the set of all convex combinations of two r-tuples of polynomials. As an illustration, consider the following r-tuple of real polynomials

$$\underline{N}_\mu = (K_{1j_1}(s),\ K_{2j_2}(s),\ \ldots,\ K_{k-1,j_{k-1}}(s),\ (1-\mu)K_{k1}(s) + \mu K_{k2}(s),\ K_{k+1,j_{k+1}}(s),\ \ldots,\ K_{rj_r}(s)). \tag{7.88}$$

Then, the line segment becomes

$$\underline{\Delta}_\mu = \underline{A}\,\underline{N}_\mu . \tag{7.89}$$

We now state without proof the following result.

Theorem 7.10 Consider the interval plant (7.87), and construct the $m4^m$ line segments, as explained above. Then, the interval plant (7.87) is stable (with respect to the left half plane), if and only if all the $m4^m$ line segments $\underline{\Delta}_\mu$ are stable.

Example 7.9 Let $G(s) = (s^3 + \alpha s^2 - 2s + \beta)/(s^4 + 2s^3 - s^2 + \gamma s + 1)$, where $\alpha \in [-1, -2]$, $\beta \in [0.5,1]$, $\gamma \in [0,1]$. We have

$K_{11}(s) = K_{12}(s) = s^3 - s^2 - 2s + 0.5$, $K_{13}(s) = K_{14}(s) = s^3 - 2s^2 - 2s + 1$

$K_{21}(s) = K_{23}(s) = s^4 + 2s^3 - s^2 + 1$, $K_{22}(s) = K_{24}(s) = s^4 + 2s^3 - s^2 + s + 1$.

In order to check that a given controller C(s) stabillizes the entire family of plants, we only need to check that it stabilizes the following four segments:

$$\underline{N}_{\mu 1} = ((1-\mu)K_{11}(s) + \mu K_{13}(s),\ K_{21}(s)) = (s^3 - (1+\mu)s^2 - 2s + 0.5(1+\mu),\ \ s^4 + 2s^3 - s^2 + 1)$$

$$\underline{N}_{\mu 2} = ((1-\mu)K_{11}(s) + \mu K_{13}(s),\ K_{22}(s)) = (s^3 - (1+\mu)s^2 - 2s + 0.5(1+\mu),\ \ s^4 + 2s^3 - s^2 + s + 1)$$

$$\underline{N}_{\mu 3} = (K_{11}(s),\ (1-\mu)K_{21}(s) + \mu K_{22}(s)) = (s^3 - s^2 - 2s + 0.5,\ \ s^4 + 2s^3 - s^2 + \mu s + 1)$$

$$\underline{N}_{\mu 4} = (K_{13}(s), (1-\mu)K_{21}(s) + \mu K_{22}(s)) = (s^3 - 2s^2 - 2s + 1,\ \ s^4 + 2s^3 - s^2 + \mu s + 1).$$

Our second generalization deals with both the uncertain parameters and the region of stability. Recall the uncertain characteristic polynomial (7.71),

$$\Delta(\lambda; p) = \sum_{i=0}^{n} a_i(p)\lambda^i \quad ; \qquad p \in \mathbf{R}^m \tag{7.90}$$

$$p_i \in [p_i^-, p_i^+] \tag{7.91}$$

but unlike (7.71), $a_i(p)$ is not an arbitrary polynomial. We assume that $a_i(p)$ depend affine linearly on the parameters $p_1, p_2, \ldots, p_m$. Each of these parameters are known only within given bounds $[p_i^-, p_i^+]$. In this case, the resulting set of polynomials turns out to be a polytope.That is, this set is the convex hull of the 2^m polynomials obtained by setting p to an extreme point.

Theorem 7.11 Consider the uncertain polynomial (7.90) - (7.91), with $a_i(p)$ linear. Let $\aleph$ be a simply connected region in the complex plane. Then $\sigma(\Delta(\lambda;p)) \subset \aleph$ for all $p_i \in [p_i^-, p_i^+]$, if and only if all the $m2^{m-1}$ exposed edges (line segments) have the same property.

Remarks

1. Theorem 7.11 is more general than Theorem 7.10. However, the number of segments in the former is much bigger.

2. In both theorems, the stability test reduces to one with a single parameter μ. As we have already shown at the beginning of this section, this is a simple task.

3. In the case where the coefficients $a_i(p)$ are general polynomial functions of p, the results of this section are of no help. In such cases we go back to Theorem 7.8 .

NOTES AND REFERENCES

The concept of critical constraint was recognized as early as 1929 by Frazer and Duncan [1], for continuous systems. The discrete counterpart was developed by Jury and Pavlidis [1]. This concept was generalized by Gutman [1] to include P-transformable regions and for arbitrary (polynomial) regions by Gutman and Chojnowski [2]. In a remarkable paper, Anderson and Scott [1] use root clustering inequalities in the minimization problem (7.27) and obtain the set of m+L polynomial equalities (7.28). This set has a finite number of solutions. To reduce the number of equations in (7.28), Walach and Zeheb [2] use the concept of zero set to obtain m+1 equations. However, there is no discussion on the possibility of an infinite number of solutions. A different computational reduction was suggested by Gutman [5]. In that article the critical constraint is combined with a simple minimization problem. Once again, the possibility of an infinite number of solutions is not excluded. The approach taken in the present chapter is based on Gutman and Chojnowski [2]. It reduces the minimization problem (7.27) to a simpler one (7.31). Subsequently, the set (7.32) contains m+1 equations in m+1 unknowns and at the same time we retain the important property of a finite number of solutions. This is used in Section 7.4 to construct a compensator having the following possible properties: relative closed loop stability, proper or strictly proper, special structure, minimum or non-minimum phase, and stable compensator. This can be achieved since we work in the parameter space. The basic idea of Section 7.5 can be traced back to Levine and Athans [1]. More recent treatment can be found in Kuhn and Schmidt [1]. Following the matrix equations form for compensator design of Kabamba and Longman [1], we extend the results to any M-transformable region (Theorem 7.3). Theorem 7.4 is a generalization of the form used by Hyland and Bernstein [1], without their optimal projection. Discussion on those results and more can be found in Fischer and Gutman [1]. That paper presents the robust analysis of Section 7.6. The discussion on interval polynomials and plants presented in Section 7.7 reflects some important recent results. Algorithm 7.3 can be traced back to Gutman [6]. Theorem 7.9 is due to Kharitonov [2], while its proof is taken from Yeung and Wang [1]. Theorem 7.10, the "box theorem", is taken from Chapellat and Bhattacharyya [1] and Theorem 7.11, the "edge theorem" is due to Bartlett, Hollot and Lin [1]. Two other important contributions are Barmish [1] and Anagnost, Desoer and Minnichelli [1].

APPENDIX

Gradient Matrices

In Chapter 7 we use the gradient of a scalar with respect to a matrix. Before presenting a detailed discussion, we need an index arithmetic for matrix manipulations.

A matrix $A \in R^{n \times m}$ is denoted by A_{nm}, where n and m are free indices. If in an expression a specific index appears twice, it means a summation with respect to this index. For example, if $B \in \mathbf{R}^{m \times r}$ and $C \in \mathbf{R}^{r \times n}$ then the product BC is denoted by $[BC]_{mn} = B_{mi}C_{in}$, which means $BC = \sum_{i=1}^{r} B_{mi} C_{in}$. Another example is $\mathrm{tr}(A) = A_{ii}$, which means $\sum_{i=1}^{n} A_{ii}$. In matrix transposition, we change the indices order. That is, if $B \in \mathbf{R}^{n \times d}$ and $C \in \mathbf{R}^{n \times k}$, then $B'C = Q$ is written $B_{id}C_{ik} = Q_{dk}$. Note that since each element in index arithmetic is a scalar, we can change the order as needed. We now define the gradient of a scalar z with respect to a matrix F.

Definition A.1 Given a scalar z and a matrix F, $\frac{\partial z}{\partial F} = B$, where $B_{ij} = \frac{\partial z}{\partial F_{ij}}$.

Theorem A.1

(i) $\frac{\partial}{\partial F} \mathrm{tr}(F' MF) = 2MF$, if M symmetric,

(ii) $\frac{\partial}{\partial F} \mathrm{tr}(BFC) = B'C'$, if $B, F, C \in \mathbf{R}^{n \times n}$,

(iii) $\frac{\partial}{\partial F} \mathrm{tr}(F) = I$, if $F \in \mathbf{R}^{n \times n}$.

Proof We prove only (i). Parts (ii), (iii) are proved along the same lines. $F'MF$ is denoted by $F_{ij}M_{ik}F_{kj}$. Thus, for any a, b we write

$$\begin{aligned} \frac{\partial}{\partial F_{ab}} (F_{ij} M_{ik} F_{kj}) &= M_{ak}F_{kb} + F_{ib}M_{ia} \\ &= M_{ak}F_{kb} + M_{ia}F_{ib} \\ &= (M + M')F \\ &= 2MF . \end{aligned}$$

We now turn to more complicated derivatives.

Theorem A.2 Let $F \in \mathbf{R}^{m \times p}$, $A \in \mathbf{R}^{n \times n}$, $B \in \mathbf{R}^{n \times m}$, $C \in \mathbf{R}^{p \times n}$, and let $\Lambda \in \mathbf{R}^{n \times n}$, $P \in \mathbf{R}^{n \times n}$ be symmetric. Then,

(i) $\frac{\partial}{\partial F} \mathrm{tr}\left[(BFC)^i\right] = iB'(C'F'B')^{i-1}C'$;

(ii) $\frac{\partial}{\partial F} \mathrm{tr}\,[\Lambda(BFC)^i] = \sum_{j=0}^{i-1} B'(C'F'B')^j \Lambda (C'F'B')^{i-1-j} C'$;

(iii) $\frac{\partial}{\partial F} \mathrm{tr}\,[\Lambda(BFC)^i P] = \sum_{j=0}^{i-1} B'(C'F'B')^j P\Lambda (C'F'B')^{i-1-j} C'$;

(iv) $$\frac{\partial}{\partial F}\operatorname{tr}\,[\Lambda(BFC)^i P(C'F'B')^j] = \sum_{k=0}^{i-1} B'(C'F'B')^k \Lambda(BFC)^j P(C'F'B')^{i-1-k} C'$$

$$+ \sum_{k=0}^{j-1} B'(C'F'B')^k \Lambda(BFC)^i P(C'F'B')^{j-1-k} C';$$

(v) $$\frac{\partial}{\partial F}\operatorname{tr}\,[\Lambda(A + BFC)^i P(A' + C'F'B')^j] =$$

$$= \sum_{k=0}^{i-1} B'(A' + C'F'B')^k \Lambda(A + BFC)^j P(A' + C'F'B')^{i-1-k} C'$$

$$+ \sum_{k=0}^{j-1} B'(A' + C'F'B')^k \Lambda(A + BFC)^i P(A' + C'F'B')^{j-1-k} C'.$$

Proof We prove (v). The rest are special cases of (v). Write

$$J = \operatorname{tr}\left[\Lambda(A + BFC)^i P(A' + C'F'B')^j\right]$$

$$= \Lambda_{ab}(A_{be} + B_{bc}F_{cd}C_{de}) \cdot \ldots \cdot (A_{lo} + B_{lm}F_{mn}C_{no}) \cdot \ldots \cdot P_{qr} \cdot$$

$$\cdot (A_{ur} + C_{sr}F_{ts}B_{ut}) \cdot \ldots \cdot (A_{av} + C_{wv}F_{xw}B_{ax}) .$$

We now take the derivative of J with respect to F_{yz} and obtain an $m \times p$ matrix. Consider the following two cases. (a) F is between Λ and P, and (b) F appears following P. In case (a), we consider the $k+1$ term $(k + 1 \le i)$, $A_{lo} + B_{lm}F_{mn}C_{no}$, and take the derivative with respect to F_{yz}. The result is zero except for $m = y$, $n = z$, in which case the result is $B_{ly}\, C_{zo}$. More specifically, we obtain

$$\Lambda_{ab}(A_{be} + B_{bc}F_{cd}C_{de}) \cdot \ldots \cdot (B_{ly}C_{zo}) \cdot \ldots \cdot P_{qr} \cdot (A_{ur} + C_{sr}F_{ts}B_{ut}) \ \ldots \cdot (A_{av} + C_{wv}F_{xw}B_{ax}) .$$

Now, rearranging the terms, in cyclic order moving backwards from B_{ly}, we end at C_{zo}, and obtain

$$B'(A' + C'F'B')^k \Lambda(A + BFC)^j P(A' + C'F'B')^{i-1-k} C' .$$

Since we take a derivative for each $k = 0, 1, \ldots, i - 1$, we take the sum over k, and obtain the first sum in (v). Likewise, in case (b), we consider the $k + 1$ term $(k + 1 \le j)$, $A_{ur} + C_{sr}F_{ts}B_{ut}$, so that the derivative with respect to F_{yz}, yields $C_{zr}B_{uy}$. Rearranging terms forwards in cyclic order starting at B_{uy}, yields

$$B'(A' + C'F'B')^k \Lambda(A + BFC)^i P(A' + C'F'B')^{j-1-k} C' .$$

Taking the sum as above, we obtain the second sum in (v). This completes the proof.

BIBLIOGRAPHY

Anagnost, J.J., Desoer, C.A., and Minnichelli, R.J.

[1] Graphical stability robustness tests for linear time-invariant systems: generalizations of Kharitonov's stability theorem, ***Proceedings of the IEEE Conference on Decision and Control***, **Austin,** 1988.

Anderson, B. D. O., Bose, N. K., and **Jury, E.I.**

[1] Output feedback stabilization and related problems-solution via decision methods, ***IEEE Trans. Automat. Contr.***, Vol. AC-20, pp. 53 - 66, 1975.

Anderson, B.D.O. and **Scott, R.W.**

[1] Output feedback stabilization-solution by algebraic geometry methods, ***IEEE Proc.***, Vol. 65, pp. 849-861, 1977.

Arapostathis, A., and **Jury, E.I.**

[1] Remarks on redundance in stability criteria and a counterexample to Fuller's conjecture, ***Int. J. Contr.***, Vol. 29, pp. 1027-1034, 1979.

Astrom, K.J.

[1] ***Introduction to Stochastic Control Theory***, Academic Press, New York, 1970.

Barmish, B.R.

[1] A generalization of Kharitonov's four-polynomial concept for robust stability problems with linearly dependent coefficient perturbations, ***IEEE Trans. Automat. Contr.,*** Vol. AC-34, pp. 157-165, 1989.

Barnett, S.

[1] ***Polynomials and Linear Control Systems***, Marcel Dekker, Inc. New York, 1983.

[2] A note on matrix equations and root location, ***IEEE Trans. Automat. Contr.***, Vol. AC-20, pp. 158-159, 1975.

Barnett, S., and **Scraton, R.E.**

[1] Location of matrix eigenvalues in the complex plane, ***IEEE Trans. Automat Contr.***, Vol. AC-27, pp. 966-967, 1982.

Barnett, S., and **Storey, C.**

[1] The Lyapunov matrix equation and Schwarz's form, ***IEEE Trans. Automat. Contr.***, Vol. AC-12, pp. 117-118, 1967.

Bartlett, A.C., Hollot, C.V. and **Lin, H.**

[1] Root locations of an entire polytope of polynomials: it suffices to check the edges, ***Math. Contr. Signals Systems,*** Vol. 1, pp. 61 - 71, 1988.

Bellman, R.

[1] ***Introduction to Matrix Analysis,*** 2nd Edition, McGraw-Hill, 1970.

Bistritz, Y.

[1] Zero location with respect to the unit circle of discrete-time linear system polynomials, ***Proc. of the IEEE,*** Vol. 72, pp. 1131-1142, 1984.

Bocher, M.

[1] ***Introduction to Higher Algebra*** , Dover, New-York, 1907.

Cartan, H.

[1] ***Cours de Calcul Differentiel,*** Herman, Paris, 1979.

Chapellat, H., and **Bhattacharyya, S.P.**

[1] A generalization of Kharitonov's theorem: robust stability of interval plants, ***IEEE Trans. Automat. Contr.,*** Vol. AC-34, pp. 306-311, 1989.

Chen, C.F., and **Chan, H.W.**

[1] A note on Jury's stability test and Kalman-Bertram Lyapunov function, ***Proc. of the IEEE,*** Vol. 73, pp. 160-161, 1985.

Chen, C.F. and **Chu, H.**

[1] A matrix for evaluating Schwarz's form, ***IEEE Trans. Automat. Contr.,*** Vol. AC-11, pp. 303-305, 1966.

Chojnowski, F.

[1] Matrix root clustering criteria, ***Ph.D. Thesis,*** Technion, Dept. of Mechanical Eng., June 1986.

Chojnowski, F., and **Gutman, S.**

[1] Root clustering criteria. Part III — rational mapping approach,***IMA J. Math. Contr. Inf.,*** Vol. 6, pp. 301-307, 1989.

[2] Root clustering criteria. Part II — linear matrix equations, ***IMA J. Math. Contr. Inf.,*** Vol. 6, pp. 289-300, 1989.

Cohn, A.

[1] Ueber die Anzahl der Wurzeln einer algebraischen Gleichung in einem Kreise, ***Math. Zeit.,*** Vols. 14-15, pp. 110-148, 1914.

Fischer, M., and **Gutman, S.**

[1] Design of reduced order compensators, to appear.

Frazer, R.A., and **Duncan, W.J.**

[1] On the criteria for the stability of small motion, ***Proc. Roy. Soc. A.,*** Vol. 124, p. 642, 1929.

Fuller, A.T.

[1] Conditions for a matrix to have only characteristic roots with negative real parts, ***J. Math. Anal. Appl.,*** Vol. 23(1), pp. 71-98, 1968.

[2] On redundance in aperiodicity criteria, ***J. Math. Anal. Appl.,*** Vol. 68, pp. 371-394, 1979.

Gutman, S.

[1] Root clustering of a real matrix in an algebraic region, ***Int. J. Control,*** Vol. 29, pp. 871-880, 1979.

[2] Root clustering of a complex matrix in an algebraic region, ***IEEE Trans. Automat. Contr.,*** Vol. AC-24, pp. 647-650, 1979.

[3] A test for root clustering transformability, ***IEEE Trans. Automat. Contr.,*** Vol. AC-27, pp. 979-981, 1982.

[4] Admissible regions for root clustering, ***IMA J. Math. Contr. Inf.,*** Vol. 3, pp. 21-27, 1986.

[5] Output feedback root clustering in parameter space, ***Israel J. of Technology,*** Vol. 21, pp. 81-84, 1983.

[6] Relative stability gain in multivariable feedback systems, ***Int. J. Control,*** Vol. 40, pp. 667 - 671, 1984.

Gutman, S., and **Chojnowski, F.**

[1] Root clustering criteria. Part I - composite matrix approach, ***IMA J. Math. Contr. Inf.,*** Vol. 6, pp. 275-288, 1989.

[2] Fixed and minimal order compensators, submitted for publication.

Gutman, S. and **Jury, E.I.**

[1] A general theory for matrix root-clustering in subregions of the complex plane, ***IEEE Trnas. Automat. Contr.,*** Vol. AC-26, pp. 853-863, 1981.

Gutman, S., and **Taub, H.**

[1] Linear matrix equations and root clustering, ***Int. J. Control,*** in press.

Hermite, C.

[1] On the number of roots of an algebraic equation contained between given limits, ***J. Reine Angew. Math.,*** Vol. 52, pp. 39-51, 1854; also in ***Int. J. Contr.,*** Vol. 26, pp. 183-195, 1977 (P.C. Parks, Trans.).

Howland, J.L.

[1] Matrix equations and the separation of matrix eigenvalues, ***J. Math. Anal. Appl.,*** Vol. 33, pp. 683-691, 1971.

Hurwitz, A.

[1] Ueber die Bedingungen unter welchen eine Gleichung nur Wurzeln mit negativen reellen Teilen besitzt, ***Math. Ann.,*** Vol. 46, pp. 273-284, 1895.

Hyland, D.C., and **Bernstein, D.S.**

[1] The optimal projection equations for fixed-order dynamic compensation, ***IEEE Trans. Automat. Contrl.,*** Vol. AC-29, pp. 1034-1037, 1984.

Jury, E.I.

[1] ***Inners and Stability of Dynamic Systems,*** 2nd edition, R.E. Krieger Publishing Company, Florida, 1982.

[2] ***Theory and Application of the Z Transform Method,*** Wiley, New-Nork, 1964.

[3] Inners approach to some problems of system theory, ***IEEE Trans. Automat. Contr.,*** Vol. AC-16, pp. 233-240, 1971.

Jury, E.I., and **Ahn, S.M.**

[1] Remarks on the root clustering of a polynomial in a certain region in the complex plane, ***Quart. Appl. Math.,*** pp. 203-205, July 1974.

Jury, E.I., and **Pavlidis, T.**

[1] Stability and aperiodicity constraints for system design, ***IEEE Prof. Group of Circuit Theory,*** Vol. CT-10, pp. 137-141, 1963.

Kabamba, P.T., and **Longman, R.W.**

[1] An integrated approach to reduced-order control theory, ***Optimal Contr. Appl. Methods***, Vol. 4, pp. 405-415, 1983.

Kailath, T.

[1] ***Linear Systems,*** Prentice-Hall, Englewood Cliffs, N.J., 1980.

Kalman, R.E.

[1] Algebraic characterization of polynomials whose zeros lie in a certain algebraic domain, ***Proc. Nat. Acad. Sci.,*** Vol. 64, pp. 818-823, 1969.

Kalman, R.E., and **Bertram, J.E.**

[1] Control system analysis and design via the second method of Lyapunov, I. Continuous time systems, Discrete time systems, ***ASME Trans., J. Basic Eng.,*** pp. 371-400, 1960.

Kharitonov, V.L.

[1] Distribution of the roots of the characteristic polynomial of an autonomous system, ***Avtomatica i Telemekhanica,*** Vol. 5, pp. 42-47, 1981.

[2] Asymptotic stability of an equilibrium position of a family of systems of linear differential equations, ***Differential Equations,*** Vol. 14, pp. 1483 - 1485, 1979.

Kuhn, U., and **Schmidt, G.**

[1] Fresh look into the design and computation of optimal output feedback controls for linear multivariable systems, ***Int. J. Control,*** Vol. 46, pp. 75-95, 1987.

Lancaster, P., and **Tismenetsky, M.**

[1] ***Theory of Matrices,*** Academic Press, Orlando, 1985.

Levine, W.S., and **Athans, M.**

[1] On the determination of the optimal constant output feedback gains for linear multi-variable systems, ***IEEE Trans. Automat. Contr.,*** Vol. AC-15, pp. 44-48, 1970.

Lienard, A., and **Chipart, M.H.**

[1] Sur la signe de la partie reelle des racines d'une equation algebraique, ***J. Math. Pure Appl.,*** Series b, Vol. 10, pp. 291-346, 1914.

Lyapunov, M.A.

[1] Obshchaya zadacha ob ustoichivosti dvisheniya, ***Math. Soc. Kharkov,*** 1892. Translation: ***Probleme generale de la stabilite du mouvement,*** Princeton Univ. Press, N.J. 1949.

MacDuffee, C.C.

[1] ***The Theory of Matrices,*** Chelsea, New-York, 1933.

Marden, M.

[1] ***Geometry of Polynomials,*** 2nd ed., Amer. Math. Soc., 1966.

Mazko, A.G.

[1] Generalization of Lyapunov's theorem for the class of regions bounded by algebraic curves, ***Soviet Autom. Control,*** pp. 93-96, 1982.

Oppenheim, A. Inequalities connected with definite Hermitian forms, ***J. London Math. Soc.,*** Vol. V, pp. 114-119, 1930.

Parks, P.C.

[1] A new proof of the Routh-Hurwitz stability criterion using the second method of Lyapunov, ***Proc. Comb. Philos. Soc.,*** Vol. 58, pp. 694-702, 1962.

[2] Comment on "the frequency domain solution of regulator problems", ***IEEE Trans. Automat. Contr.,*** Vol. AC-11, p. 334, 1966.

[3] Lyapunov and Schur-Cohn stabilty criterion, ***IEEE Trans. Automat. Cont.,*** Vol. AC-9, p. 121, 1964.

[4] A new proof of Hermite's stability criterion and a generalization of Orlando's formula, ***Int. J. Control,*** vol. 26, pp. 197-206, 1977.

Routh, E.J.

[1] ***Stability of a Given State of Motion,*** London: MacMillan, 1877.

Schneider, H.

[1] Positive operators and an inertia theorem, ***Numerische Mathematik,*** Vol. 7, pp. 11 - 17, 1965.

Schur, I.

[1] Ueber Potenzreihen, die im Innern des Einheitskreises beschraenkt sind, ***J, Math.,*** Vol. 147, pp. 205-232, 1917.

[2] Bemerkungen sur Theorie der beschraenkten Bilinearformen mit unendlich vielen Veraenderlichen, ***J. Math.,*** Vol. 140, pp. 1-28, 1911.

Schwarz, H.R.

[1] Ein Verfahren zur Stabilitaetsfrage bes Matrizen-Eigenwerte Probleme, ***Z. Angew. Math. Phys.,*** Vol. 7, pp. 473-500, 1956.

Sondergeld, K.P.

[1] A generalization of the Routh-Hurwitz stability criteria and an application to a problem in robust controller design, ***IEEE Trans. Automat. Contr.,*** Vol. AC-28, pp. 965-970, 1983.

Stein, P.

[1] Some general theorems on iterants, ***J. Res. Natl. Bur. Stan.,*** Vol. 48, pp. 82-83, 1952.

Stephanos, C.

[1] Sur une extension du calcul des substitutions lineaires, ***J. Math. Pures Appl.,*** Vol. 6, pp. 73-128, 1900.

Tarski, A.

[1] ***A Decision Method for Elementary Algebra,*** University of California Press, Berkeley, 1951.

Taub, H. and **Gutman, S.**

[1] Roots of composite polynomials — an application to root clustering, ***Linear Alg. Appl.,*** Vol. 87, pp. 181-188, 1987.

[2] A symmetric matrix criterion for polynomial root clustering, ***IEEE Trans. Circuits and Systems,*** To appear.

Walach, E. and **Zeheb, E.**

[1] Root distribution for the ellipse, ***IEEE Trans. Automat. Contr.,*** Vol. AC-27, pp. 960-963, 1982.

[2] Generalized zero set of multiparameter polynomials and feedback stabilization, ***IEEE Trans. Circuits and Systems,*** Vol. CAS-29, pp. 15-23, 1982.

Waring, E.

[1] Problems, ***Philos. Trans. Roy. Soc.,*** Vol. 53, pp. 294-299, 1763.

Yeung, .S., and **Wang, S.S.**

[1] A simple proof of Kharitonov's theorem, ***IEEE Trans. Automat. Contr.,*** Vol. AC-32, pp. 822-823, 1987.

Zeheb, E. and **Hertz, D.**

[1] Complete root distribution with respect to parabolas and some results with respect to hyperbolas and sectors, ***Int. J. Control,*** vol. 36, pp. 517-530, 1982.

Lecture Notes in Control and Information Sciences

Edited by M. Thoma and A. Wyner

Vol. 97: I. Lasiecka/R. Triggiani (Eds.)
Control Problems for Systems
Described by Partial Differential Equations
and Applications
Proceedings of the IFIP-WG 7.2
Working Conference
Gainesville, Florida, February 3-6, 1986
VIII, 400 pages, 1987.

Vol. 98: A. Aloneftis
Stochastic Adaptive Control
Results and Simulation
XII, 120 pages, 1987.

Vol. 99: S. P. Bhattacharyya
Robust Stabilization Against
Structured Perturbations
IX, 172 pages, 1987.

Vol. 100: J. P. Zolésio (Editor)
Boundary Control and Boundary Variations
Proceedings of the IFIP WG 7.2 Conference
Nice, France, June 10-13, 1987
IV, 398 pages, 1988.

Vol. 101: P. E. Crouch,
A. J. van der Schaft
Variational and Hamiltonian
Control Systems
IV, 121 pages, 1987.

Vol. 102: F. Kappel, K. Kunisch,
W. Schappacher (Eds.)
Distributed Parameter Systems
Proceedings of the 3rd International Conference
Vorau, Styria, July 6–12, 1986
VII, 343 pages, 1987.

Vol. 103: P. Varaiya, A. B. Kurzhanski (Eds.)
Discrete Event Systems:
Models and Applications
IIASA Conference
Sopron, Hungary, August 3-7, 1987
IX, 282 pages, 1988.

Vol. 104: J. S. Freudenberg/D. P. Looze
Frequency Domain Properties of Scalar
and Multivariable Feedback Systems
VIII, 281 pages, 1988.

Vol. 105: Ch. I. Byrnes/A. Kurzhanski (Eds.)
Modelling and Adaptive Control
Proceedings of the IIASA Conference
Sopron, Hungary, July 1986
V, 379 pages, 1988.

Vol. 106: R. R. Mohler (Editor)
Nonlinear Time Series and
Signal Processing
V, 143 pages, 1988.

Vol. 107: Y. T. Tsay, L.-S. Shieh, St. Barnett
Structural Analysis and Design
of Multivariable Systems
An Algebraic Approach
VIII, 208 pages, 1988.

Vol. 108: K. J. Reinschke
Multivariable Control
A Graph-theoretic Approach
274 pages, 1988.

Vol. 109: M. Vukobratović/R. Stojić
Modern Aircraft Flight Control
VI, 288 pages, 1988.

Vol. 110: B. J. Daiuto, T. T. Hartley,
S. P. Chicatelli
The Hyperbolic Map and Applications
to the Linear Quadratic Regulator
VI, 114 pages, 1989

Vol. 111: A. Bensoussan, J. L. Lions (Eds.)
Analysis and Optimization
of Systems
XIV, 1175 pages, 1988.

Vol. 112: Vojislav Kecman
State-Space Models of Lumped
and Distributed Systems
IX, 280 pages, 1988

Vol. 113: M. Iri, K. Yajima (Eds.)
System Modelling and Optimization
Proceedings of the 13th IFIP Conference
Tokyo, Japan, Aug. 31 – Sept. 4, 1987
IX, 787 pages, 1988.

Vol. 114: A. Bermúdez (Editor)
Control of Partial Differential Equations
Proceedings of the IFIP WG 7.2
Working Conference
Santiago de Compostela, Spain, July 6–9, 1987
IX, 318 pages, 1989

Vol. 115: H.J. Zwart
Geometric Theory for Infinite
Dimensional Systems
VIII, 156 pages, 1989.

Vol. 116: M.D. Mesarovic, Y. Takahara
Abstract Systems Theory
VIII, 439 pages, 1989

Lecture Notes in Control and Information Sciences

Edited by M. Thoma and A. Wyner

Vol. 117: K.J. Hunt
Stochastic Optimal Control Theory
with Application in Self-Tuning Control
X, 308 pages, 1989.

Vol. 118: L. Dai
Singular Control Systems
IX, 332 pages, 1989

Vol. 119: T. Başar, P. Bernhard
Differential Games and Applications
VII, 201 pages, 1989

Vol. 120: L. Trave, A. Titli, A. M. Tarras
Large Scale Systems:
Decentralization, Structure Constraints
and Fixed Modes
XIV, 384 pages, 1989

Vol. 121: A. Blaquière (Editor)
Modeling and Control of Systems
in Engineering, Quantum Mechanics,
Economics and Biosciences
Proceedings of the Bellman Continuum
Workshop 1988, June 13–14, Sophia Antipolis, France
XXVI, 519 pages, 1989

Vol. 122: J. Descusse, M. Fliess, A. Isidori,
D. Leborgne (Eds.)
New Trends in Nonlinear Control Theory
Proceedings of an International
Conference on Nonlinear Systems,
Nantes, France, June 13–17, 1988
VIII, 528 pages, 1989

Vol. 123: C. W. de Silva, A. G. J. MacFarlane
Knowledge-Based Control with
Application to Robots
X, 196 pages, 1989

Vol. 124: A. A. Bahnasawi, M. S. Mahmoud
Control of Partially-Known
Dynamical Systems
XI, 228 pages, 1989

Vol. 125: J. Simon (Ed.)
Control of Boundaries and Stabilization
Proceedings of the IFIP WG 7.2 Conference
Clermont Ferrand, France, June 20–23, 1988
IX, 266 pages, 1989

Vol. 126: N. Christopeit, K. Helmes
M. Kohlmann (Eds.)
Stochastic Differential Systems
Proceedings of the 4th Bad Honnef Conference
June 20–24, 1988
IX, 342 pages, 1989

Vol.127: C. Heij
Deterministic Identification
of Dynamical Systems
VI, 292 pages, 1989

Vol. 128: G. Einarsson, T. Ericson,
I. Ingemarsson, R. Johannesson,
K. Zigangirov, C.-E. Sundberg
Topics in Coding Theory
VII, 176 pages, 1989

Vol. 129: W. A.Porter, S. C. Kak (Eds.)
Advances in Communications and
Signal Processing
VI, 376 pages, 1989.

Vol. 130: W. A. Porter, S. C. Kak,
J. L. Aravena (Eds.)
Advances in Computing and Control
VI, 367 pages, 1989

Vol. 131: S. M. Joshi
Control of Large Flexible Space Structures
IX, 196 pages, 1989.

Vol. 132: W.-Y. Ng
Interactive Multi-Objective Programming
as a Framework for Computer-Aided Control
System Design
XV, 182 pages, 1989.

Vol. 133: R. P. Leland
Stochastic Models for Laser Propagation
in Atmospheric Turbulence
VII, 145 pages, 1989.

Vol. 134: X. J. Zhang
Auxiliary Signal Design in Fault
Detection and Diagnosis
XII, 213 pages, 1989

Vol. 135: H. Nijmeijer, J. M. Schumacher (Eds.)
Three Decades of Mathematical System Theory
A Collection of Surveys at the Occasion of the
50th Birthday of Jan C. Willems
VI, 562 pages, 1989

Vol. 136: J. Zabczyk (Ed.)
Stochastic Systems and Optimization
Proceedings of the 6th IFIP WG 7.1
Working Conference,
Warsaw, Poland, September 12-16, 1988
VI, 374 pages. 1989

Lecture Notes in Control and Information Sciences

Edited by M. Thoma and A. Wyner

Vol. 137: S. L. Shah, G. Dumont (Eds.)
Adaptive Control Strategies for Industrial Use
Proceedings of a Workshop
Kananaskis, Canada, 1988
VI, 360 pages. 1989

Vol. 138: D. C. McFarlane, K. Glover
Robust Controller Design Using Normalized Coprime Factor Plant Descriptions
X, 206 pages. 1990

Vol. 139: V. Hayward, O. Khatib (Eds.)
Experimental Robotics I
The First International Symposium
Montreal, June 19-21, 1989
XVII, 613 pages. 1990

Vol. 140: Z. Gajić, D. Petkovski, X. Shen
Singulary Perturbed and Weakly Coupled Linear Control Systems
A Recursive Approach
VII, 202 pages. 1990

Vol. 141: S. Gutman
Root Clustering in Parameter Space
VIII, 153 pages. 1990